electricity
two

Hayden Electricity One–Seven Series

Harry Mileaf, Editor-in-Chief

electricity one Producing Electricity □ Atomic Theory □ Electrical Charges □ Electron Theory □ Current □ Voltage □ Magnetism □ Electromagnetism

electricity two D-C Circuits □ Direct Current □ Resistors □ Ohm's Law □ Power □ Series Circuits □ Parallel Circuits □ Series-Parallel Circuits □ Kirchhoff's Laws □ Superposition □ Thevenin's Theorem □ Norton's Theorem

electricity three A-C Circuits □ Alternating Current □ A-C Waveforms □ Resistive Circuits □ Inductors □ Inductive Circuits □ Transformers □ Capacitors □ Capacitive Circuits

electricity four LCR Circuits □ Vectors □ RL Circuits □ RC Circuits □ LC Circuits □ Series-Parallel Circuits □ Resonant Circuits □ Filters

electricity five Test Equipment □ Meter Movements □ Ammeters □ Voltmeters □ Ohmmeters □ Wattmeters □ Multimeters □ Vacuum-Tube Voltmeters

electricity six Power Sources □ Primary Cells □ Batteries □ Photo, Thermo, Solar Cells □ D-C Generators □ A-C Generators □ Motor-Generators □ Dynamotors

electricity seven Electric Motors □ D-C Motors □ A-C Motors □ Synchronous Motors □ Induction Motors □ Reluctance Motors □ Hysteresis Motors □ Repulsion Motors □ Universal Motors □ Starters □ Controllers

electricity two

HARRY MILEAF EDITOR-IN-CHIEF

revised second edition

HAYDEN BOOKS
A Division of Howard W. Sams & Company
4300 West 62nd Street
Indianapolis, Indiana 46268 USA

© 1966 and 1976 by Hayden Books
A Division of Howard W. Sams & Co.

SECOND EDITION
FIFTEENTH PRINTING—1989

International Standard Book Number: 0-8104-5946-9
Library of Congress Catalog Card Number: 75-45504

Printed in the United States of America

preface

This volume is one of a series designed specifically to teach electricity. The series is logically organized to fit the learning process. Each volume covers a given area of knowledge, which in itself is complete, but also prepares the student for the ensuing volumes. Within each volume, the topics are taught in incremental steps and each topic treatment prepares the student for the next topic. Only *one* discrete topic or concept is examined on a page, and *each* page carries an illustration that graphically depicts the topic being covered. As a result of this treatment, neither the text nor the illustrations are relied on solely as a teaching medium for any given topic. Both are given for *every* topic, so that the illustrations not only complement but reinforce the text. In addition, to further aid the student in retaining what he has learned, the important points are summarized in text form on the illustration. This unique treatment allows the book to be used as a convenient review text. Color is used not for decorative purposes, but to accent important points and make the illustrations meaningful.

In keeping with good teaching practice, all technical terms are defined at their point of introduction so that the student can proceed with confidence. And, to facilitate matters for both the student and the teacher, key words for each topic are made conspicuous by the use of italics. Major points covered in prior topics are often reiterated in later topics for purposes of retention. This allows not only the smooth transition from topic to topic, but the reinforcement of prior knowledge just before the declining point of one's memory curve. At the end of each group of topics comprising a lesson, a summary of the facts is given, together with an appropriate set of review questions, so that the student himself can determine how well he is learning as he proceeds through the book.

Much of the credit for the development of this series belongs to various members of the excellent team of authors, editors, and technical consultants assembled by the publisher. Special acknowledgment of the contributions of the following individuals is most appropriate: Frank T. Egan, Jack Greenfield, and Warren W. Yates, principal contributors; Peter J. Zurita, Steven Barbash, Solomon Flam, and A. Victor Schwarz, of the publisher's staff; Paul J. Barotta, Director of the Union Technical Institute; Albert J. Marcarelli, Technical Director of the Connecticut School of Electronics; Howard Bierman, Editor of *Electronic Design;* E. E. Grazda, Editorial Director of *Electronic Design;* and Irving Lopatin, Editorial Director of the Hayden Book Companies.

HARRY MILEAF
Editor-in-Chief

contents

using electricity

In itself, electricity is nothing more than an interesting phenomenon. To be of practical use, it must be made to perform some work or function. Generally, this requires that the electricity be controlled, and often converted to other forms of energy. The physical means for accomplishing this transition from phenomenon to practical use is the *electric circuit.*

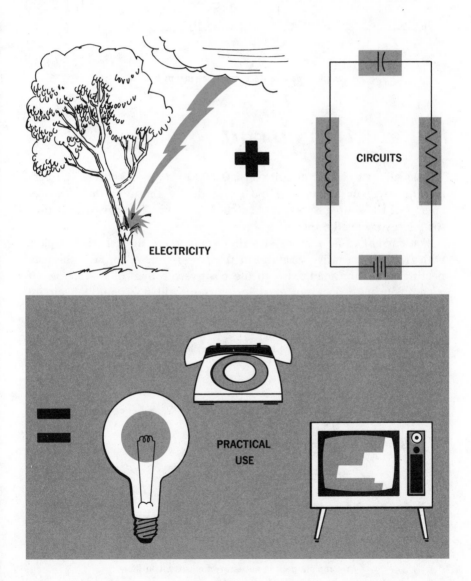

ELECTRICITY

CIRCUITS

PRACTICAL USE

**Three elements are necessary
to have an electric circuit**

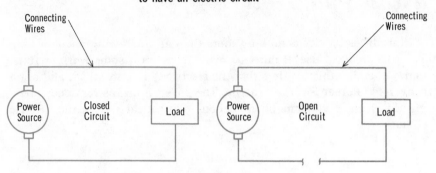

A complete path is necessary for current to flow

the electric circuit

Basically, an electric circuit consists of (1) a power source; (2) connecting wires, or conductors; and (3) a device that uses the electrical energy of the source to accomplish some purpose. The device that uses the energy is called the *load*.

For current to flow in an electric circuit, there must be a complete path from the negative terminal of the power source, through the connecting wires and load, back to the positive terminal of the source. If a complete path does not exist, no current will flow, and the circuit is called an *open circuit*.

**The load is a device that
uses electrical energy**

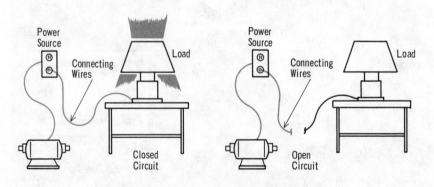

A complete path is necessary for current to flow

the switch

An electric circuit has to provide a complete path for current flow only when electrical energy is needed by the load. At all other times, the circuit is kept "open," and no current flows.

All switches perform the same basic function of opening and closing electric circuits

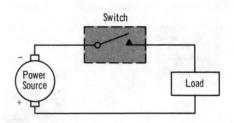

The opening and closing of an electric circuit is normally accomplished by a *switch*. In its simplest form, a switch consists of two pieces of conducting metal that are connected to the circuit wires. These two pieces of metal are arranged so that they may easily be made to either touch each other or be separated. When they touch, a complete path for current flow exists and the circuit is closed. When they are separated, no current can flow, and the circuit is open.

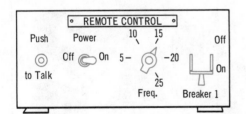

Many types of electrical switches are used today. Each type has its own schematic representation, so that you can look at a circuit diagram and know what type of switch is being used

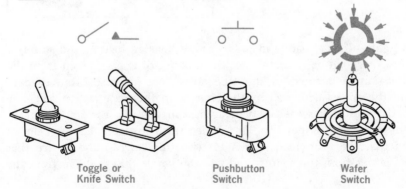

Toggle or Knife Switch

Pushbutton Switch

Wafer Switch

the load

In a simple electric circuit, the load is the device that takes the electrical energy from the power source and uses it to perform some useful function. To do this, the load may convert the electrical energy to another form of energy, such as light, heat, or sound, or it may merely change or control the amount of energy delivered by the source.

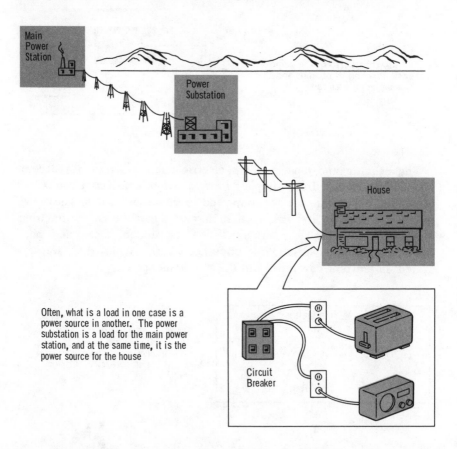

Often, what is a load in one case is a power source in another. The power substation is a load for the main power station, and at the same time, it is the power source for the house

A light bulb is a load, and so is a motor, toaster, heater, and so forth. The type of load used determines the amount of energy taken from the power source. Because of this, the term "load" is also often used to mean the *power* delivered by the source. In this case, when someone says the load is increased or decreased, it means that the source is supplying more or less power. (This is described in detail later.) Keep in mind that the term load stands for two things: (1) the *device* that takes power from the source, and (2) the *power* that is taken from the source.

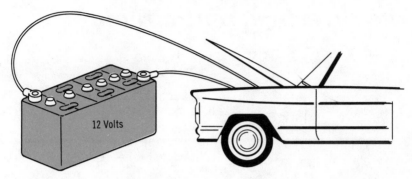

One of the most widely used voltage sources today
is the battery

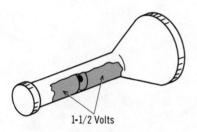

There is a wide variety of sizes, shapes,
and types of batteries in use today

the power source

The power source produces electrical energy by chemical, magnetic, or other means. (This is explained in Volume 1.) This energy is usually in the form of a *difference of electrical potential* between the output terminals of the source, which is called an *electromotive force*. Usually, the abbreviation emf is used in place of the term electromotive force. Emf is measured in volts, and so the source producing it is called a *voltage source*. The polarity of the voltage source determines in which direction the current flows in the circuit; and the amount of voltage supplied by the source determines how much current will flow. Power sources are covered in detail in Volume 6.

the direct-current circuit

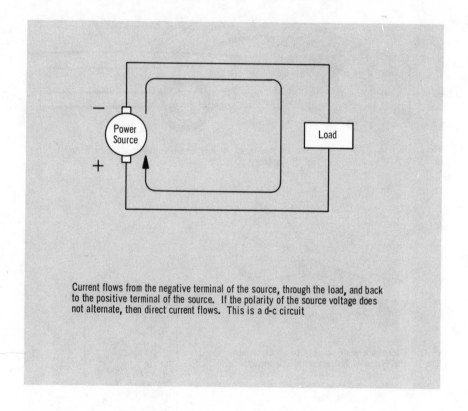

Current flows from the negative terminal of the source, through the load, and back to the positive terminal of the source. If the polarity of the source voltage does not alternate, then direct current flows. This is a d-c circuit

Since electron current always flows out of the negative terminal of the power source, the current flow in a circuit will always be in the same direction, if the polarity of the source voltage always remains the same. This type of current flow is called direct current, and the source is called a direct-current source. Any circuit that uses a direct-current source is then a direct-current circuit. For simplicity, direct current is generally abbreviated as dc, and we then speak of d-c sources, d-c voltage, d-c current, and d-c circuits. The three types of sources most often used for d-c circuits are the battery, the d-c generator, and the electronic power supply. Regardless of the type of d-c source used, the theory of operation of all d-c circuits is the same. D-c theory is covered in this volume.

When the voltage polarity of the power source changes, or alternates, the direction of current flow will also alternate. This type of current is called *alternating current* (ac), and is covered in Volume 3.

what controls current flow?

Generally, electric circuits are designed for some specific amount of current flow. If too little current flows in the circuit, the load will not operate properly, or maybe not at all. If too much current flows, the voltage source or load could be damaged. There are only two factors that determine how much current will flow in a d-c circuit. The first is the amount of voltage that is supplied by the power source, and the second is how well the load and the wires conduct current.

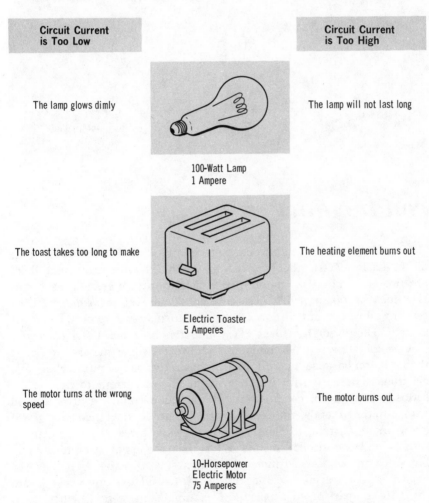

Circuit Current is Too Low

Circuit Current is Too High

The lamp glows dimly

The lamp will not last long

100-Watt Lamp
1 Ampere

The toast takes too long to make

The heating element burns out

Electric Toaster
5 Amperes

The motor turns at the wrong speed

The motor burns out

10-Horsepower
Electric Motor
75 Amperes

The current flowing in any circuit must be controlled if the circuit is to work properly

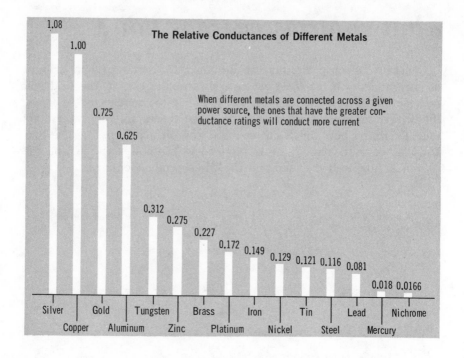

The Relative Conductances of Different Metals

When different metals are connected across a given power source, the ones that have the greater conductance ratings will conduct more current

Silver 1.08
Copper 1.00
Gold 0.725
Aluminum 0.625
Tungsten 0.312
Zinc 0.275
Brass 0.227
Platinum 0.172
Iron 0.149
Nickel 0.129
Tin 0.121
Steel 0.116
Lead 0.081
Mercury 0.018
Nichrome 0.0166

conductance

Not all materials conduct current equally well. If you recall some of the basic electric theory given in Volume 1, you know that there are two basic kinds of material in which we are interested, as far as electricity is concerned. These are *conductors* and *insulators*. Conductors allow current to flow easily, and insulators oppose the flow of current. The reason for this is that conductors have many *free electrons*.

Most metals are good conductors. However, some metals are better than others because not all metals have the same number of free electrons. The ease with which a metal allows current to flow is described by the term *conductance*. If the same voltage source is used with different metals, the metals with a high conductance rating will allow more current to flow. The bar graph gives the conductance ratings of some similar metals. Silver has the greatest conductance; but since copper is used more often than silver because it is cheaper, copper is given a conductance rating of 1, and the other metals are rated in comparison to copper. For example, tungsten, which is used in light bulbs, has only 0.312 the conductance of copper. Therefore, copper will allow more than 3 times as much current to flow than will tungsten if they were both connected across the same power source.

summary

☐ The electric circuit is the physical means for accomplishing the transition of electricity from phenomenon to practical use. ☐ The electric circuit consists of (1) a power source, (2) connecting wires or conductors, and (3) a device that uses the electrical energy of the source. ☐ The device that uses the electrical energy is the load. ☐ For current flow, there must be a complete path from the negative terminal of the power source, through the connecting wires and load, back to the positive terminal of the source. ☐ A switch is used to open and close the path between the source and the load. An open switch means that there is no continuous path through the circuit; the circuit is therefore open. A closed switch means that there is a continuous path; the circuit is then closed.

☐ The power source produces electrical energy by chemical, magnetic, or other means. ☐ The power source discussed in this volume is a direct-current source: a battery, a d-c generator, or an electronic power supply. The theory of circuit operation is the same regardless of the source type used. ☐ The circuit is usually designed to operate with a specific current flow. Too low a value may cause the load to be inoperative; too high a value may damage the voltage source or load. ☐ Two factors determine how much current will flow in a d-c circuit: (1) the amount of voltage supplied, and (2) how well the load and the wires conduct current.

☐ Two materials of interest in electricity are conductors and insulators. Conductors allow current to flow; insulators oppose it. ☐ The ease with which a metal allows current to flow is described as conductance. ☐ All metals are given a conductance rating which shows how well they conduct electricity, as compared to copper. ☐ The best conductor is silver, but copper is used more often because it is cheaper.

review questions

1. What are the three basic elements of an electric circuit?
2. What is meant by an *open* circuit?
3. What controls a circuit so that it is open or closed?
4. What is meant by a *load*?
5. Can a battery ever be considered a load? Explain.
6. What do a battery, d-c generator, and electronic power supply have in common?
7. What controls the current flow in a circuit?
8. What is meant by *conductance rating*?
9. Conductance ratings are based on what metal?
10. What can happen if there is too much current in a circuit?

resistance

The term *conductance* is used to describe how well a material *allows* current to flow. Another way to look at this is that the *low-conductance* materials oppose or *resist* the flow of electric current. Some materials, then, offer more *resistance* to the flow of electrons than other materials. This is actually the way that materials are rated in the field of electricity.

If you were to cut a piece of each of the more common metals to a standard size, and then connect the pieces to a battery, one at a time, you would find that different amounts of current would flow. Each metal offers a different resistance to the movement of electrons.

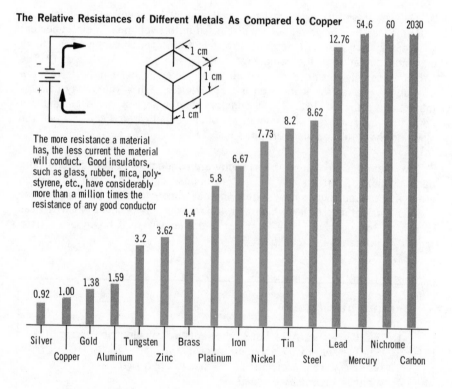

The Relative Resistances of Different Metals As Compared to Copper

The more resistance a material has, the less current the material will conduct. Good insulators, such as glass, rubber, mica, polystyrene, etc., have considerably more than a million times the resistance of any good conductor

Silver Copper Gold Aluminum Tungsten Zinc Brass Platinum Iron Nickel Tin Steel Lead Mercury Nichrome Carbon

0.92 1.00 1.38 1.59 3.2 3.62 4.4 5.8 6.67 7.73 8.2 8.62 12.76 54.6 60 2030

The standard size that is usually used to test the resistance of metals is a 1-centimeter cube. The bar graph shows the resistance of some common metals as compared to copper. Silver is a better conductor than copper because it has less resistance. Nichrome has 60 times more resistance than copper, and so copper will conduct 60 times as much current as nichrome if they were connected to the same battery, one at a time.

Point of
Current
Measurement

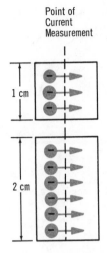

1 cm

2 cm

When a conductor is made thicker, it
will conduct more current and will
have less resistance

Conductors with greater cross-sectional
area have more free electrons available,
and, therefore, have less resistance

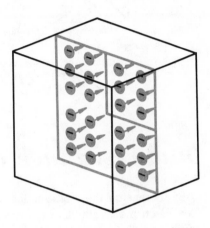

how resistance can be decreased

Actually, the resistance of any material depends on the number of free electrons the material has. If you think back to what was explained in Volume 1, you should recall that electric current is rated in amperes; and 1 ampere is 6,280,000,000,000,000,000 free electrons passing any given point in a wire in 1 second. Therefore, a good conductor must have a number of free electrons available to allow several amperes to flow. Since current is a measurement of electrons passing a point in a wire, we can make more free electrons available by using a *thicker* piece of metal so that *more current* will flow.

A piece of copper 2 centimeters high and 1 centimeter wide will have twice as many free electrons available along the point at which current is being measured than a piece of copper only 1 centimeter high and 1 centimeter wide. The copper that is twice as high will conduct twice as much current. If you use copper that is 2 centimeters wide, you will double the current and halve the resistance again. When you increase the width or height of a piece of metal, you are increasing its *cross-sectional area*. The *greater* the cross-sectional *area* of a wire, the *lower* its *resistance*.

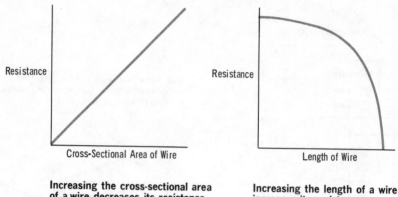

Increasing the cross-sectional area
of a wire decreases its resistance

Increasing the length of a wire
increases its resistance

how resistance can be increased

By increasing the cross-sectional area of a conductor, you make more free electrons available for current, and lower the resistance of the conductor. This might lead you to think that by changing the length of a piece of copper, you could accomplish the same thing. But this isn't so. Although a longer piece of copper has more free electrons in the whole piece, the extra free electrons are not made available along the line of current measurement. Actually, each length of conductor has a certain amount of resistance. When you add an *extra length* of copper, you are adding *more resistance*. The longer a wire is, the more resistance it has.

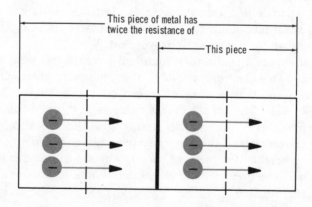

how resistance is controlled

You can see now that the resistance of a piece of wire can be increased by making it longer, or decreased by making it shorter. You can also lower the resistance by increasing its cross-section, or raise the resistance by decreasing its cross-section.

If you double the length, you will double the resistance. Because of this relationship, we say that the resistance of a wire is *directly proportional* to its *length*.

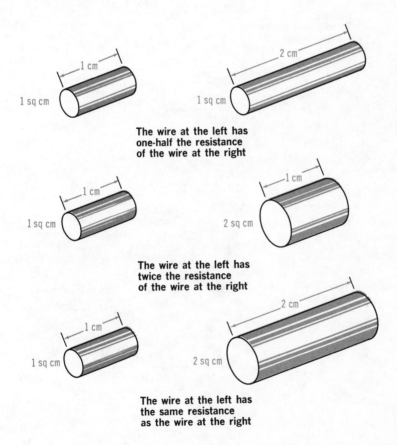

The wire at the left has
one-half the resistance
of the wire at the right

The wire at the left has
twice the resistance
of the wire at the right

The wire at the left has
the same resistance
as the wire at the right

If you double the cross-sectional area of wire, you will halve its resistance. As a result, we say that the resistance of a wire is *inversely proportional* to its *cross-section.*

Therefore, by choosing the proper metal for a conductor, and making it with a certain cross-section and length, you can produce any kind of resistance effect you want.

effect of temperature

Actually, the relative values of resistance that were given before apply to the metals when they are at about room temperature. At higher or lower temperatures, the resistances of all materials change. In most cases, when the temperature of a material goes up, its resistance goes up too. But with some other materials, increased heat causes the resistance to go down. The amount that the resistance is

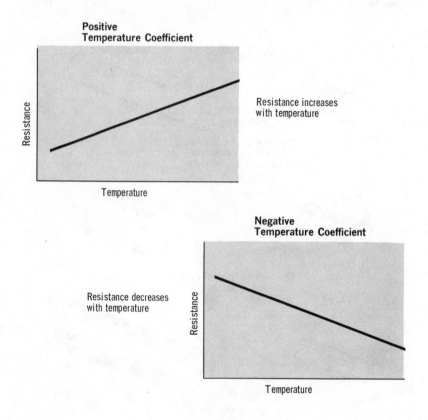

**Positive
Temperature Coefficient**

Resistance

Temperature

Resistance increases
with temperature

**Negative
Temperature Coefficient**

Resistance decreases
with temperature

Resistance

Temperature

affected by each degree of temperature change is called the *temperature coefficient*. And the words positive and negative are used to show whether the resistance goes up or down with temperature.

When a material's resistance goes *up* as temperature is increased, it has a *positive* temperature coefficient.

When a material's resistance goes *down* as temperature is increased, it has a *negative* temperature coefficient.

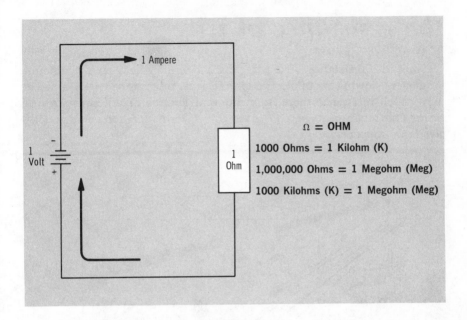

the unit of resistance

During the early 1800's, a German scientist named Georg Simon Ohm performed a number of experiments with electricity, and made some of the first important discoveries about the nature of electrical resistance. In his honor, the unit of resistance is called the *ohm*.

A conductor is said to have a resistance of 1 ohm when an emf of 1 volt causes a current of 1 ampere to flow through that conductor. If, of course, 1 volt causes only ½ ampere to flow, then the conductor has twice the resistance, or 2 ohms. By using this relationship, the exact resistance of all types, sizes, and shapes of conductors can be found. Resistance can vary from fractions of an ohm to kilohms (1000 ohms) and megohms (1,000,000 ohms). The Greek letter omega (Ω) is often used as the symbol for the ohm.

CONVERSION OF UNITS
ohms ÷ 1000 = kilohms (K)
ohms ÷ 1,000,000 = megohms (Meg)
kilohms (K) ÷ 1000 = megohms (Meg)
kilohms × 1000 = ohms
megohms × 1,000,000 = ohms
megohms × 1000 = kilohms (K)
500,000 ohms = 500 kilohms = 0.5 megohm
or
500,000 Ω = 500 K = 0.5 Meg

the resistance of wire

Since the resistance of a wire has a definite effect on the amount of current flowing in an electric circuit, it is important for you to know how much resistance there is in different lengths of different sizes of wire. The table on page 2-17 gives this information for commercially available copper wire.

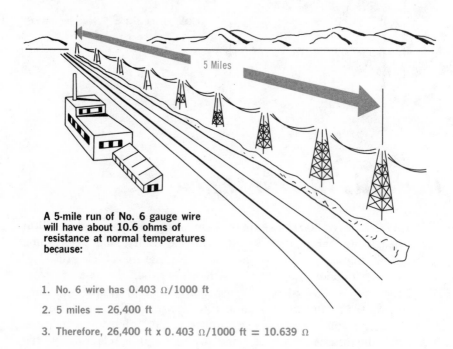

5 Miles

A 5-mile run of No. 6 gauge wire will have about 10.6 ohms of resistance at normal temperatures because:

1. No. 6 wire has 0.403 Ω/1000 ft
2. 5 miles = 26,400 ft
3. Therefore, 26,400 ft x 0.403 Ω/1000 ft = 10.639 Ω

The table was originally called the Brown and Sharpe Wire Gauge, but is now known as the American Standard Wire Gauge. The different diameters of copper wire are identified by gauge numbers. There are 40 gauges shown, ranging from 0000 through 36. The higher the gauge number, the smaller the diameter and cross-sectional area of the wire. Therefore, the *higher* numbered *gauges* of wire have *more resistance*. The table gives the diameter for each gauge and the typical resistance for 1000 feet of wire. Resistances are given for both the normal temperature of 70°F, and a high temperature of 167°F.

the resistance of wire (cont.)

AMERICAN STANDARD WIRE GAUGES

Dimensions and Typical Resistances of Commercial Copper Wire

B & S Gauge No.	Diameter of Bare Wire (Inches)	Ohms per 1000 ft		Current Capacity (Amperes)	
		70°F	167°F	Rubber Insulation	Other Insulation
0000 (4/0)	0.460	0.050	0.060	160-248	193-510
000 (3/0)	0.410	0.062	0.075	138-215	166-429
00 (2/0)	0.365	0.080	0.095	120-185	145-372
0	0.325	0.100	0.119	105-160	127-325
1	0.289	0.127	0.150	91-136	110-280
2	0.258	0.159	0.190	80-118	96-241
3	0.229	0.202	0.240	69-101	83-211
4	0.204	0.254	0.302	60-87	72-180
5	0.182	0.319	0.381	52-76	63-158
6	0.162	0.403	0.480	45-65	54-134
7	0.144	0.510	0.606	45-65	54-134
8	0.128	0.645	0.764	35-48	41-100
9	0.114	0.813	0.963	35-48	41-100
10	0.102	1.02	1.216	25-35	31-75
11	0.091	1.29	1.532	25-35	31-75
12	0.081	1.62	1.931	20-26	23-57
13	0.072	2.04	2.436	20-26	23-57
14	0.064	2.57	3.071	15-20	18-43
15	0.057	3.24	3.873	15-20	18-43
16	0.051	4.10	4.884	6	10
17	0.045	5.15	6.158	6	10
18	0.040	6.51	7.765	3	6
19	0.036	8.21	9.792	3	6
20	0.032	10.3	12.35	–	–
21	0.028	13.0	15.57	–	–
22	0.025	16.5	19.63	–	–
23	0.024	20.7	24.76	–	–
24	0.020	26.2	31.22	–	–
25	0.018	33.0	39.36	–	–
26	0.016	41.8	49.64	–	–
27	0.014	52.4	62.59	–	–
28	0.013	66.6	78.93	–	–
29	0.011	82.8	99.52	–	–
30	0.010	106	125.50	–	–
31	0.009	134	158.20	–	–
32	0.008	165	199.50	–	–
33	0.007	210	251.60	–	–
34	0.006	266	317.30	–	–
35	0.005	337	400.00	–	–
36	0.005	423	504.50	–	–

how much current can wire carry?

The resistance of a wire determines how much current will flow through the wire when it is connected across a voltage source. But there are two other factors that should also be considered when it comes to current flow. One is that although the wire resistance will *allow* a certain amount of current to flow, no more current can flow than the power source is able to supply. This is why a wire is rarely put directly across a source. Sources have a *safe limit* of maximum current they can supply before they *burn out*. This is explained in detail in Volume 6.

Because electric current heats a wire, current is used as a heat source in the electric blanket

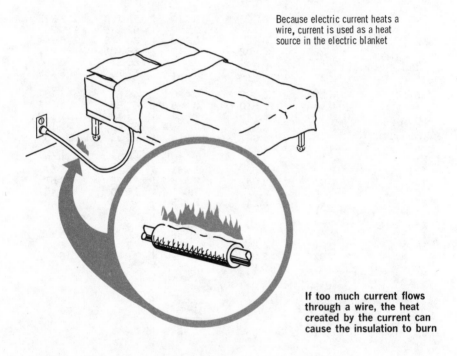

If too much current flows through a wire, the heat created by the current can cause the insulation to burn

The second factor to consider is how much current the wire can *safely* carry. You should remember from what was explained in Volume 1 that electric *current heats* a wire. If the wire becomes too hot, its insulation covering will burn, and the wire can deteriorate. This is the reason why the AWS table on page 2-17 gives the maximum range of current that each gauge of wire can safely carry. As the table shows, if another insulation material which is better than rubber is used, the wire can carry more current.

the resistance of the load and of the power source

Up to this point, you have studied the resistance of ordinary wires, and how they affect current flow. But circuits must have a load if they are to perform a function with electricity. In most cases, the resistance of the load is much greater than the resistance of the circuit wiring. For example, a motor uses a large number of turns of wire, so the wiring of a motor is actually very long. And a heater, toaster, and electric iron use high-resistance conductors to produce heat. Therefore, the load resistance has more effect on how much current flows than the resistance of the circuit wires.

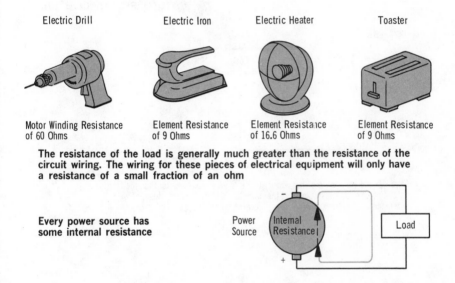

Electric Drill   Electric Iron   Electric Heater   Toaster

Motor Winding Resistance   Element Resistance   Element Resistance   Element Resistance
of 60 Ohms   of 9 Ohms   of 16.6 Ohms   of 9 Ohms

The resistance of the load is generally much greater than the resistance of the circuit wiring. The wiring for these pieces of electrical equipment will only have a resistance of a small fraction of an ohm

Every power source has some internal resistance

Power Source — Internal Resistance — Load

Some quantity of resistance is present in everything that conducts current. When current flows in a circuit, electrons leave the negative terminal of the power source, go around the wires and through the load, and reenter the power source through the positive terminal. The power source does not continually provide electron current from the negative terminal without replenishing its supply of electrons. The current that enters the positive terminal of the source must be conducted through the source to complete the circuit. Therefore, current flows from positive to negative *inside the source*. And every source has some *internal resistance* that also opposes current flow. This internal resistance is usually very low compared to the load.

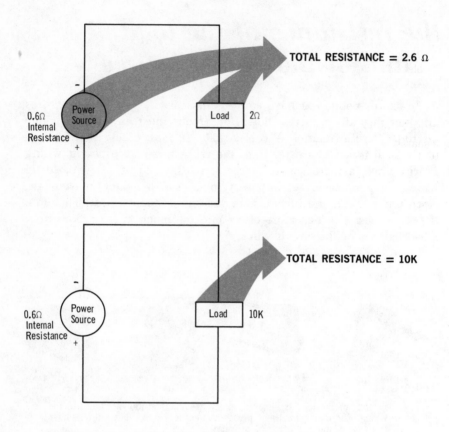

If the internal resistance of the power source is very much smaller than the load resistance, the total resistance of the circuit is that of the load alone. The resistance of long transmission lines, however, must be considered

total circuit resistance

The total resistance of a circuit is the sum of the individual resistances of the power source, the circuit wiring, and the load. As you know, the load resistance is generally much higher than the resistance of either the source or the wiring. In such cases, you can ignore the smaller resistances and consider the load resistance as the total resistance. But if the load resistance is small too, then those resistances should be added to get the true total resistance. Very rarely does the circuit wiring have enough resistance to be considered. But with long electric power, telephone, or telegraph lines, the resistance is large enough so that it must be considered.

summary

☐ Materials are rated in the field of electricity by their resistance to the flow of electrons. ☐ Good conductors have a low resistance, and good insulators have a high resistance. ☐ The resistance of a wire is determined by the type of material and the dimensions of the wire. The greater the cross-sectional area of the wire, the lower is its resistance. The longer the wire is, the greater is its resistance. In other words, the resistance of a wire is inversely proportional to its cross-sectional area and directly proportional to its length.

☐ The temperature coefficient of a material describes the effect that temperature has on the resistance of the material. A positive temperature coefficient means the resistance increases with an increase in temperature; a negative coefficient means the resistance decreases with an increase in temperature. ☐ The unit of resistance is the ohm. 1 ohm exists when an emf of 1 volt causes a current of 1 ampere to flow.

☐ The American Standard Wire Gauge lists resistance and current capacity characteristics for different sizes of copper wire. The higher the AWG number, the smaller is the wire diameter and therefore the greater is the resistance. ☐ Although a particular wire gauge can allow a high current flow, the actual current in a circuit is determined by: (1) the safe limit the source can supply before burning out, and (2) the safe limit the wire can carry. ☐ The total resistance of a circuit is the sum of the individual resistances of the power source, the circuit wiring, and the load. ☐ Usually, the load resistance is so much greater than that of the source or the wiring that it can be considered the total resistance.

review questions

1. Would a metal with a conductance rating of 0.99 be a good or poor insulator?
2. A certain material has a resistance of 15 ohms. If its cross-sectional area were tripled, what would be its resistance?
3. Define *temperature coefficient*.
4. Does the length of a wire affect its temperature coefficient?
5. Define *internal resistance* of a power source.
6. Does copper have a positive or a negative temperature coefficient?
7. What must be considered when calculating total circuit resistance?
8. What is the unit and symbol for resistance?
9. What is the American Standard Wire Gauge?
10. If a copper wire is heated, what happens to its resistance?

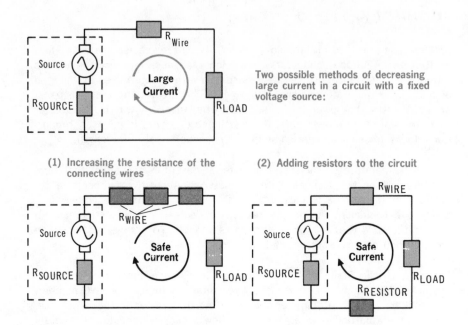

Two possible methods of decreasing large current in a circuit with a fixed voltage source:

(1) Increasing the resistance of the connecting wires

(2) Adding resistors to the circuit

resistors

Very often, if a load is connected to an existing source of voltage, too much current might flow in the circuit. This could happen if the resistance of the load was very low or the output voltage of the source was very high. The current could be decreased by reducing the source voltage, but usually this is impossible, or at least impractical. As you already know, the only other way to decrease the current is to add resistance to the circuit. This might be done by increasing the resistance of the voltage source, the load, or the connecting wires. However, the resistances of the source and the load are built right into these devices and cannot be changed. This leaves the connecting wires; but their resistance is so low that miles and miles of wire might be needed just to add a few hundred ohms to a circuit. Of course, connecting wires with a higher resistance could be used, and as a matter of fact, this has been done in the past for certain applications. However, if this was always done, it would greatly increase the number of different types of wires used to interconnect electric circuits. What is needed, then, is a method of easily adding various amounts of resistance to a circuit without drastically changing either its physical size or the materials used to make it. *Resistors* are the electric circuit components used to accomplish this.

how resistors are used

Resistors are used for *adding* resistance to an electric circuit. Basic-ally, they are materials that offer a high resistance to current flow. The materials most often used for resistors are *carbon*, and special *metal alloys*, such as nichrome, constantan, and manganan. A resistor is connected into a circuit in such a way that the circuit current flows through it as well as through the load and the source. The *total circuit resistance* is then the sum of the individual resistances of the *load*, the *source*, the connecting *wires*, and the *resistor*. You can see from this that by just adding the proper resistor to the circuit, the resistance of the circuit can be changed to almost any value.

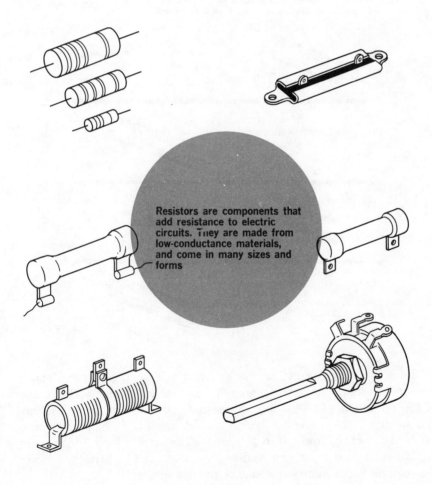

Resistors are components that add resistance to electric circuits. They are made from low-conductance materials, and come in many sizes and forms

tolerance

The basic characteristic of any resistor is the number of ohms of resistance it has. This is called the *resistor value,* and is normally marked on the resistor in some manner. However, the value marked on a resistor is really only a "nominal" value. The actual value may be somewhat higher or lower. The reason for this is that resistors are normally mass produced. And, as with all mass produced items, variations occur during manufacture. To account for these variations, resistors are marked with a *tolerance.*

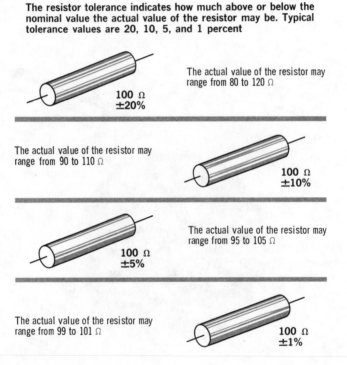

The resistor tolerance indicates how much above or below the nominal value the actual value of the resistor may be. Typical tolerance values are 20, 10, 5, and 1 percent

The actual value of the resistor may range from 80 to 120 Ω

100 Ω
±20%

The actual value of the resistor may range from 90 to 110 Ω

100 Ω
±10%

The actual value of the resistor may range from 95 to 105 Ω

100 Ω
±5%

The actual value of the resistor may range from 99 to 101 Ω

100 Ω
±1%

The resistor tolerance is usually given as a percentage, and indicates how much higher or lower than the nominal value the actual value of the resistor *might* be. Thus, a resistor marked as being 100 ohms and having a 10 percent tolerance would actually have a resistance value of anywhere from 90 to 110 ohms. The most common tolerances are 20, 10, 5, and 1 percent. And as you can probably guess, the lower the tolerance, the more expensive is the resistor.

current rating

You already know that when current flows in a wire it generates heat in the wire. The cause of this is the wire's resistance. The higher this resistance is, the more heat is produced. In a resistor, the resistance is concentrated in a small area, and so the heat generated by current flow is also concentrated in a small area. As a result, a resistor can become very hot when connected in a circuit. This means that it must be able to either withstand the heat generated, or give up the heat to the surrounding air. If it can do neither, it will eventually be damaged or destroyed. Even if the heat is not severe enough to damage the resistor, it may cause a large *change* in the *resistance,* since, as you remember, the resistance of all materials changes as their temperatures change.

Every resistor has a *maximum current rating,* and should not be used in a circuit where more than this maximum current will flow. Otherwise, the resistor may burn out. The current rating of a resistor is normally given as the *wattage rating* of the resistor. This will be explained later.

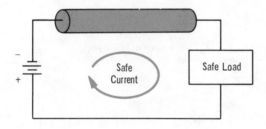

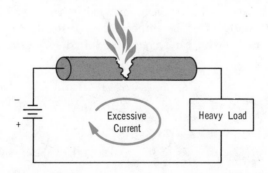

All resistors are designed for a certain maximum current. If this current is exceeded, the heat generated in the resistor will cause a change in the resistance and possibly even destroy the resistor. The wattage rating determines the maximum current a resistor can carry

resistor types

Based on what you now know about resistors, you may think that choosing a resistor for a circuit is a simple matter. It is just a matter of picking one that has the right resistance value and proper tolerance, and which also can carry the circuit current without burning out.

Although these are important considerations, they are not the only ones. There are many more, such as the cost of the resistor, how sturdy or rugged it is, how it is mounted in a circuit, and whether or not age or long use would cause changes in its resistance value. You can see, then, that you must consider many things when you pick a resistor. However, not all of these things are important in every case.

ALL-PURPOSE RESISTOR

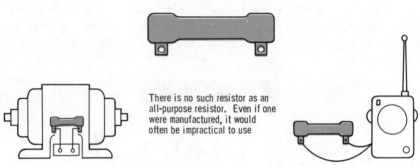

There is no such resistor as an all-purpose resistor. Even if one were manufactured, it would often be impractical to use

For example, the cost of the resistors in a small table radio was very important to the manufacturer, but the tolerances were not. If a single type of resistor was made that could be used in all circuits and under all conditions, it would be very expensive and would have many features that would often be unnecessary. Instead, different types of resistors are made, with each type best suited for certain uses.

Composition Resistor   Wirewound Resistor   Film Resistor

Most of the resistors used today are one of two basic types: *composition resistors,* or *wirewound resistors.* However, a third type, called *film resistors,* is being used more and more.

composition resistors

In most cases where a resistor is used, the requirements are not severe, and what is needed is a resistor that will do the job as cheaply as possible. The type of resistor most often used in these cases is the *composition resistor*. The most common type of composition resistor consists essentially of a *powdered-carbon* resistance element, a tubular *plastic case* for sealing and protecting the resistance element, and *wire leads* for connecting the resistor into a circuit. As you can see from the bar graph on page 2-10, carbon has a resistance 2030 times that of copper. It only takes small amounts of carbon, therefore, to obtain high resistance. The powdered carbon is mixed with an insulating material, called a *binder,* and the value of the resistor depends on the relative amounts of carbon and binder material used.

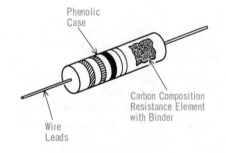

Phenolic
Case

Carbon Composition
Resistance Element
with Binder

Wire
Leads

Always be careful when soldering composition resistors. Heat from the soldering iron may be conducted along the leads to the resistance element and cause a significant change in the resistor's value

Advantages	Disadvantages
Small	Overheats with high currents
Rugged	Large temperature coefficient
Low Cost	Wide tolerance range

Composition resistors are made with resistance values of from less than 10 ohms to more than 20 million ohms (20 Megs), and with tolerances of 20, 10, and 5 percent. They cannot carry high currents without overheating, and have large temperature coefficients. They have the advantages, however, of small size, ruggedness, and low cost. In general, composition resistors are used in applications where large currents are not involved and narrow tolerances not required.

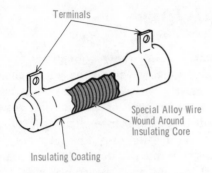

Precision-type wirewound resistors have very small tolerances and keep within these
tolerances during use. This
property of maintaining a
small tolerance under all
operating conditions is called
the STABILITY of a
resistor

Power-type wirewound resistors can carry high
currents and dissipate large amounts of heat

wirewound resistors

The two main shortcomings of composition resistors are their limited current-carrying ability, and the difficulty in making them with small tolerances. Both of these limitations can be overcome, but at an increase in cost, by using resistance elements of special resistance wire instead of powdered carbon. Long lengths of wire are usually needed to obtain the necessary resistance values; to keep the resistor as small as possible, the wire is wound around a core. Resistors made in this way are called *wirewound resistors*.

There are two basic types of wirewound resistors: the *power* type, and the *precision* type. The power type is used for circuits having large currents, whereas the precision type is used when resistances with very small tolerances are required. Essentially, both types are made by winding special alloy *wire* around an insulating *core,* and then applying a ceramic, plastic, or other insulating *coating*. The ends of the winding are attached to metal caps at each end of the core. The caps have some form of *terminal* for connecting the resistor to a circuit. The high currents used with power type resistors generate large amounts of heat, which must be transferred to the surrounding air or dissipated. These resistors are therefore often large, since the more surface area a body has, the more heat it can transfer. The power-type wirewound resistors are made with resistance values of a few ohms to thousands (K) of ohms, with tolerances of 10 or 20 percent. Precision-type wirewound resistors are made with low resistance values, as low as 0.1 ohm and with tolerances as small as 0.1 percent. To obtain such small tolerances, costly materials and construction methods are used, and as a result, precision-type resistors are very expensive.

film resistors

Film resistors can be considered a compromise between composition resistors and precision-type wirewound resistors. They have some of the *accuracy* and *stability* of the wirewound type, but are smaller, more rugged, and cheaper.

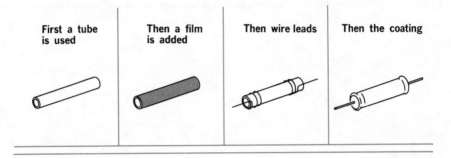

First a tube is used   **Then a film is added**   **Then wire leads**   **Then the coating**

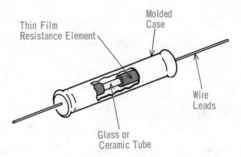

Thin Film Resistance Element

Molded Case

Wire Leads

Glass or Ceramic Tube

Film resistors are named according to the film material used. Thus, there are carbon film resistors, boron-carbon film resistors, metal film resistors, and metallic oxide film resistors

Film resistors are usually made by depositing, by a special process, a *thin film* of resistance material on a glass or ceramic tube. Leads for connecting the resistor into a circuit are connected to caps at the end of the tube. An insulating coating is then molded around the unit for protection.

The resistance of a film resistor is determined by the material used for the film, and its *thickness*. In general, the thicknesses used range between 0.00001 and 0.00000001 inch. You can see why these resistors are often called *thin film resistors*.

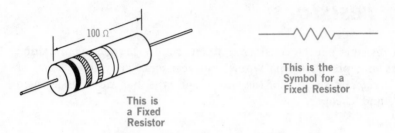

This is
a Fixed
Resistor

This is the
Symbol for a
Fixed Resistor

The value of a fixed resistor is set and cannot be varied

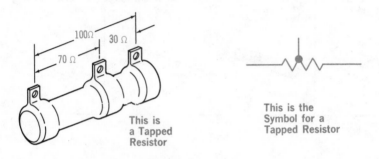

This is
a Tapped
Resistor

This is the
Symbol for a
Tapped Resistor

You can obtain more than one resistance value from a tapped
resistor. However, each of these values is fixed

fixed resistors

You have now learned how resistors are classified according to the
materials used for their resistance elements. But there is another way
of classifying resistors. This is by whether their resistance value is
fixed and unchangeable, or whether it can be *varied*. The types of
resistors described earlier have two leads, each connected to one end
of the resistance element; when these resistors are connected into a
circuit, their entire resistance value is added to the circuit. You can
see, therefore, that a *fixed* resistor has only one resistance value. However, there is a special type of fixed resistor that has more than one
value. This type has, in addition to the terminals at the ends of the
resistance element, one or more other terminals somewhere between
the ends of the resistance element. By connecting different terminals
into a circuit, different values of resistance can be obtained. Each of
these different resistances, however, is still in itself a fixed resistance.
This type of resistor is called a *tapped* resistor.

Fixed resistors can be either of the composition, wirewound, or film
type.

adjustable resistors

From the preceding pages, you can see that a fixed resistor has no flexibility as far as its value of resistance is concerned. It has one value, which cannot be changed or varied. The tapped resistor offers some flexibility, since more than one value of resistance can be obtained from it. However, the number of resistance values you can obtain from a tapped resistor is usually limited to three or four. What is desirable in many applications is a resistor from which you can obtain a *range of resistance values,* from zero up to some maximum; for example, a resistor that you could adjust for any value from 0 to 100 ohms, or maybe 0 to 25K. One type of resistor that gives this flexibility is the *adjustable resistor.* An adjustable resistor is similar to a tapped, fixed wirewound resistor, except that all or part of the winding is exposed. A *movable clamp,* with a terminal attached, makes contact with the winding and can be moved to any position along the length of the winding. The resistance between the movable terminal and either of the end terminals then depends on the position of the movable clamp.

These resistors are not built to be frequently adjusted. They are normally set to the required resistance value when they are installed in a circuit, and then left at this value.

This is the Symbol for an Adjustable Resistor

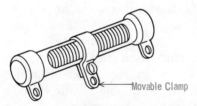

Movable Clamp

This is an Adjustable Resistor

With an adjustable resistor, you can obtain any resistance value covered by the range of the resistor

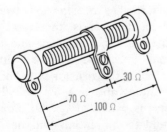

30 Ω

70 Ω

100 Ω

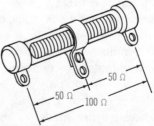

50 Ω

50 Ω

100 Ω

Though a wide range of resistance values is possible, these resistors are made in such a way that frequent adjustment is impractical

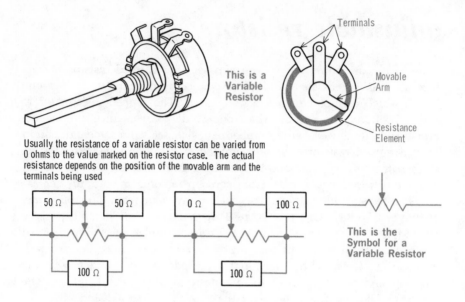

This is a
Variable
Resistor

Terminals

Movable
Arm

Resistance
Element

Usually the resistance of a variable resistor can be varied from
0 ohms to the value marked on the resistor case. The actual
resistance depends on the position of the movable arm and the
terminals being used

This is the
Symbol for a
Variable Resistor

variable resistors

In many electrical devices a resistance value must be changed frequently. Examples of this are the volume control on your radio, the brightness control on your television set, or an electric light dimmer or motor speed control. This cannot be done using an adjustable resistor, since it would be difficult and time consuming. The resistors used must be continuously variable over a certain range of resistances, the same as adjustable resistors, but they must also be very *easy* to vary and built to withstand *frequent* adjustments. Resistors which do this are called *variable* resistors. Usually, a variable resistor consists of a circular resistance element contained in a housing or case. The element can be wirewound, composition, or film. A movable contact can slide across the element while maintaining electrical contact with it.

The movable contact is turned by a shaft. The resistance between movable contact and the ends of the element depends, therefore, on the position of the shaft. Both ends of the resistance element and the movable contact are connected to external terminals. When all three of these terminals are connected into a circuit, the resistor is called a *potentiometer*. When only the center terminal and one of the other terminals are used in a circuit, the resistor is called a *rheostat*. Sometimes, rheostats are made without the end terminal that will not be used. So remember: potentiometers and rheostats are both variable resistors. The only difference is the way they are used in a circuit.

resistor color code

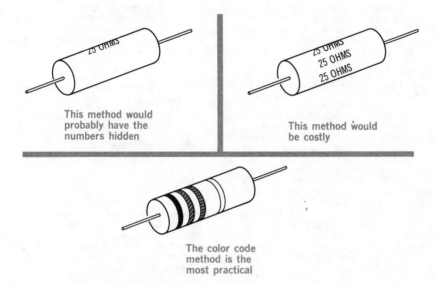

This method would probably have the numbers hidden

This method would be costly

The color code method is the most practical

It is not always practical to mark the resistance value using numbers on axial-lead, fixed composition resistors. Instead, colored bands are used. The relationship of these bands to the resistance value is called the resistor color code

All resistors have their resistance value marked on them in some way. At first, you might suppose that this would always be done using numbers; for example, 50 ohms or 1000 ohms. The larger power resistors, precision resistors, and variable resistors are marked this way, but this is impractical for small fixed, composition resistors. These resistors are often too small to be marked that way. Also, they are tubular in shape and have axial leads, and can therefore be physically mounted in a circuit in any position. If their resistance values were marked in numbers, there would be a good chance the numbers would be hidden once the resistors were connected in a circuit. Of course, the numbers could be marked all around the resistor, but this would be difficult and costly from a manufacturing standpoint. This problem has been solved by using a series of *colored bands* around the resistors to indicate their resistance values. The positions of the bands and their color, making up what is called a *color code,* indicate the resistance values. A single standard color code has been adopted by the United States Armed Forces and the Electronic Industries Association (EIA) for fixed composition, axial-lead resistors.

standard resistor color code

First Significant Figure: The color of the first band indicates the first digit of the resistor value. For example, using the Color Code Table, if this band is yellow, the first digit is 4.

Multiplying Value: The color of the third band indicates how much the first two digits are to be multiplied to obtain the resistance value. For example, using the Color Code Table, if this band is green, the first two digits are multiplied by 100,000. This band can also be looked on as indicating the number of zeros to be added after the second digit. When used this way, the number of zeros shown in the Significant Figures column of the Color Code Table is the number of zeros to add. For example, if this band is orange, add three zeros after the second digit. But if the band is black, no zeros are added. If the third band is gold or silver, the multiplication factor must be used.

Second Significant Figure: The color of the second band indicates the second digit of the resistor value. For example, using the Color Code Table, if this band is black, the second digit is 0.

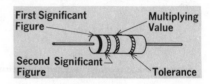

Tolerance: The color of the fourth band indicates the tolerance of the resistor. For example, if this band is gold, the resistor tolerance is ±5 percent. If there is no tolerance band on a resistor, the tolerance is automatically ±20 percent.

COLOR CODE TABLE

Color	Significant Figures	Multiplying Value	Tolerance
Black	0	1	—
Brown	1	10	—
Red	2	100	—
Orange	3	1000	—
Yellow	4	10,000	—
Green	5	100,000	—
Blue	6	1,000,000	—
Violet	7	10,000,000	—
Gray	8	100,000,000	—
White	9	1,000,000,000	—
Gold	—	0.1	±5%
Silver	—	0.01	±10%
No Color	—	—	±20%

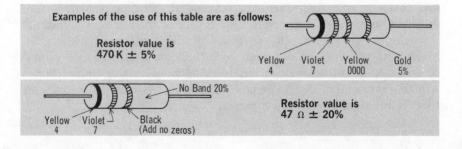

Examples of the use of this table are as follows:

Resistor value is
470 K ± 5%

Yellow Violet Yellow Gold
  4   7   0000   5%

No Band 20%

Yellow Violet Black
  4   7   (Add no zeros)

Resistor value is
47 Ω ± 20%

summary

☐ Resistors are inserted into a circuit to lower the current flow by adding resistance. ☐ There are two basic types of resistors: composition and wirewound. Film resistors, a third type, are now becoming more and more popular. ☐ Composition resistors are usually made with a powdered carbon resistance element. They have high temperature coefficients, low current capacity, and large tolerances. Their advantages include ruggedness, small size, and low cost.

☐ Wirewound resistors are usually made from special resistance wire, wound about a core. They have large current capacity, and can be of the power or precision types. The power types have large current capacity; the precision types have very small tolerances. ☐ Film resistors are usually made by depositing a thin film of resistance material on a glass or ceramic tube. They have some of the advantages of both composition and wirewound resistors.

☐ Adjustable resistors have a movable clamp that is set to a required resistance value when installed in a circuit. They are not made for frequent adjustment. ☐ Variable resistors are similar to adjustable resistors, except that they are continuously variable over a range of resistances. ☐ If all three terminals of a variable resistor are connected in a circuit, the resistor is called a potentiometer. ☐ If only the center and one of the other terminals are connected, the resistor is called a rheostat. ☐ The resistor color code for fixed composition, axial-lead resistors indicates both the nominal value and tolerance of the resistor. The first two color bands indicate the significant figures; the third band is the multiplier; and the fourth band is the tolerance.

review questions

1. Describe the construction of the three types of fixed resistors.
2. State the advantages and disadvantages of composition resistors.
3. State the advantages and disadvantages of wirewound resistors.
4. What are the advantages of film resistors?
5. Define *nominal value* of a resistor.
6. Define *tolerance* of a resistor.
7. What is the difference between a rheostat and a potentiometer?
8. How do the *tapped, variable,* and *adjustable* resistors differ?
9. Describe the resistor color code.
10. What is meant by *significant figures,* as used in the color code?

ohm's law

As you learned earlier, since voltage causes current to flow in a closed circuit and resistance opposes the flow of current, a relationship exists between voltage, current, and resistance. This relationship was first determined in a series of experiments made by Georg Simon Ohm, who, as you remember from page 2-15, is the person for whom the unit of resistance was named.

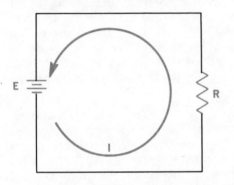

It is common practice to abbreviate current as I, voltage as E, and resistance as R

George Simon Ohm proved that current, I, in a d-c circuit is directly proportional to the voltage, E, and inversely proportional to the resistance, R

This means:
1. If you raise E, I will go up
2. If you lower E, I will go down
3. If you raise R, I will go down
4. If you lower R, I will go up

Ohm found that if the resistance in a circuit was kept constant, and the source voltage was increased, there would be a corresponding increase in current. Likewise, a decrease in voltage would cause a decrease in current. Stated another way, Ohm found that in a d-c circuit, current is directly proportional to voltage. Ohm also discovered that if the source voltage was held constant, while the circuit resistance was increased, the current would decrease. Similarly, a decrease in resistance resulted in an increase in current. In other words, current is inversely proportional to resistance. This relationship between the current, voltage, and resistance in a d-c circuit is known as Ohm's Law, and can be summarized as follows: *In a d-c circuit, the current is directly proportional to the voltage and inversely proportional to the resistance.*

equations

Strictly speaking, Ohm's Law is a statement of proportion and not a mathematical equation. However, if the current is in amperes, the voltage in volts, and the resistance in ohms, then Ohm's Law can be expressed by the equation:

$$I = E/R$$

which states that current (I) equals voltage (E) divided by resistance (R). Two variations of this equation that are very useful in analyzing d-c circuits are

$$R = E/I$$

which states that resistance (R) equals voltage (E) divided by current (I) and

$$E = IR$$

which states that voltage (E) equals current (I) times resistance (R).

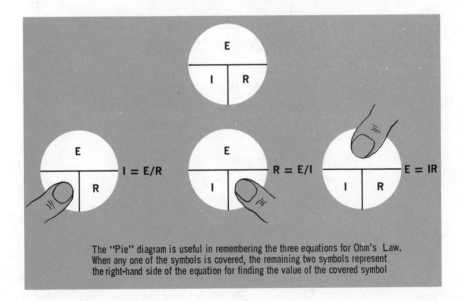

The "Pie" diagram is useful in remembering the three equations for Ohm's Law. When any one of the symbols is covered, the remaining two symbols represent the right-hand side of the equation for finding the value of the covered symbol

You can see that with these three equations, when *two* of the three circuit elements (current, voltage, and resistance) are known, the third can easily be found.

It is very important for you to *memorize* these three equations. You will be using them over and over while you study and work with circuits.

calculating current

At one time or another, you will probably have to calculate the current flowing in a circuit. You know that this can be done using Ohm's Law, so the first step is to decide which of the equations for Ohm's Law applies. A good practice to always follow at this point is to think in terms of *knowns* and *unknowns*. In any equation, the unknown is the term whose value you are trying to find. It is always the term to the left of the equal sign. The knowns are all the other terms of the equation. They are to the right of the equal sign.

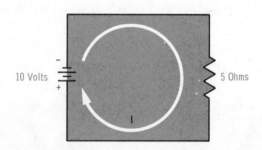

Suppose you are asked to find the current in the circuit. The first step is to study the circuit diagram and then phrase the question in your own mind as simply as possible, perhaps as the following:

How much current would an applied voltage of 10 volts cause through a resistance of 5 ohms?

Since current (I) is the unknown, you would use the equation

$$I = E/R$$
$$= 10 \text{ Volts}/5 \text{ Ohms}$$
$$= 2 \text{ Amperes}$$

In our problem, we are trying to find the value of the current, and therefore I is the unknown. As shown on page 2-37, the equation for Ohm's Law in which I is the *unknown* is

$$I = E/R$$

Therefore, this is the equation to use when calculating current in a circuit by using Ohm's Law.

calculating current (cont.)

The circuit diagram at the left shows a 20-ohm resistor used as the load in a circuit having a 100-volt battery as a voltage source. If the resistor has a maximum current rating of 8 amperes, will its rating be exceeded when the switch is closed?

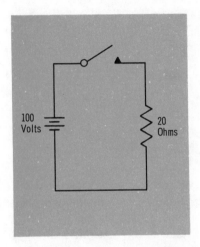
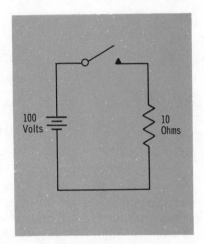

After reading the question and studying the diagram, you should see that you are really being asked two questions: (1) How much current would an applied voltage of 100 volts cause through 20 ohms of resistance, and (2) is this unknown current greater than 8 amperes? To answer the first question, the unknown is the current, and so the equation $I = E/R$ is used.

$$I = E/R = 100 \text{ volts}/20 \text{ ohms} = 5 \text{ amperes}$$

The second question can then be answered by a simple comparison. Since only 5 amperes of current flows, the 8-ampere current rating of the resistor is not exceeded.

What if a 10-ohm resistor, also with a maximum rating of 8 amperes, is used instead?

The equation $I = E/R$ is used again.

$$I = E/R = 100 \text{ volts}/10 \text{ ohms} = 10 \text{ amperes}$$

The resulting current of 10 amperes does exceed the 8-ampere current rating. The resistor will probably burn out.

calculating resistance

You calculate resistance by Ohm's Law by the equation:

$$R = E/I$$

You would use this equation to choose the proper size resistor to be connected into a circuit, or to determine the resistance of a resistor or other load already in a circuit.

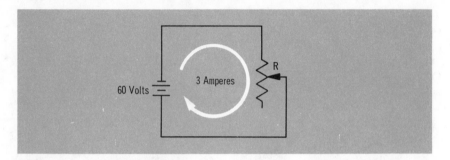

In the circuit diagram, 3 amperes of current flow in the circuit when the rheostat is set at the middle of its range. How much resistance is being added to the circuit?

What is actually being asked here is: What is the resistance through which an applied voltage of 60 volts will cause a current of 3 amperes to flow? Since resistance is the unknown, the equation R = E/I is used.

$$R = E/I = 60 \text{ volts}/3 \text{ amperes} = 20 \text{ ohms}$$

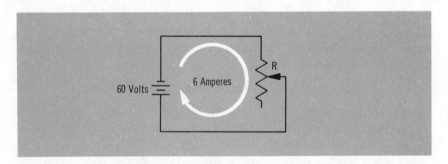

In the circuit, how much resistance would the rheostat have to add to the circuit to increase the current to 6 amperes? Again, resistance is the unknown, and the equation R = E/I is used.

$$R = E/I = 60 \text{ volts}/6 \text{ amperes} = 10 \text{ ohms}$$

Thus, to double the current, the resistance must be halved.

calculating voltage

You calculate voltage by Ohm's Law using the equation:

$$E = IR$$

If the light bulb in the circuit diagram has a resistance of 100 ohms and 1 ampere of current flows in the circuit when the switch is closed, what is the voltage output of the battery?

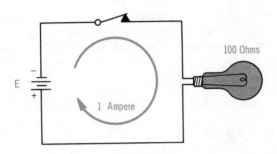

After studying the circuit diagram, you can see that what is being asked is: How much voltage will cause a current of 1 ampere to flow through a resistance of 100 ohms? Voltage is the unknown, so the equation $E = IR$ is used.

$$E = IR = 1 \text{ ampere} \times 100 \text{ ohms} = 100 \text{ volts}$$

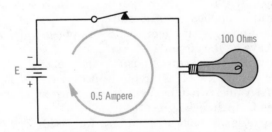

If the battery shown in the circuit wears out so that only 0.5 ampere flows in the circuit, what is the battery output voltage?

Again the voltage equation, $E = IR$, is used:

$$E = IR = 0.5 \text{ ampere} \times 100 \text{ ohms} = 50 \text{ volts}$$

You can see from this that the current was cut in half when the source voltage was reduced by one half.

summary

☐ Ohm's Law describes the relationship between voltage, current, and resistance in a d-c circuit. The Law states: In a d-c circuit, the current is directly proportional to the voltage and inversely proportional to the resistance.

☐ Ohm's Law can be expressed mathematically by three equations: $I = E/R$; this equation is used when current I is the unknown quantity in the circuit. ☐ $R = E/I$; this equation is used when resistance R is the unknown quantity in the circuit. ☐ $E = IR$; this equation is used when voltage E is the unknown quantity in the circuit.

☐ A Pie diagram is useful in remembering any of the three equations for Ohm's Law. When any one of the symbols is covered, the remaining two symbols represent the right-hand side of the equation (the known quantities). The symbol covered is the left-hand side of the equation (the unknown quantity).

review questions

1. State Ohm's Law and give its three equations.
2. Draw a circuit with a 15-volt battery power source and a 5-ohm resistive load. What is the current flowing in the circuit?
3. In what two ways can the current in Question 2 be doubled?
4. If the resistance of a circuit is increased to four times its original value, what would have to be done to the source voltage to maintain the original current flow?
5. A 100-ohm *film* resistor has a maximum current rating of 2 amperes. What is the maximum value of source voltage that can be applied to the circuit?
6. In Question 5, what effect would an applied voltage of 500 volts have on the circuit?
7. What is a *Pie diagram*? Illustrate its use.
8. If the resistance of a circuit is decreased to 1/4 of its original value, what happens to the current if the source voltage is unchanged?
9. If the resistance of a circuit is decreased to 1/4 of its original value, what circuit change can be made to maintain the original circuit current?
10. Doubling the resistance of a circuit has what effect on the current, if the source voltage is held constant? Halving the source voltage has what effect on the current, if the resistance is held constant? Doubling both the source voltage and resistance has what effect on the current?

power

As you now know, the purpose of the power source in an electric circuit is to supply electrical energy to the load. The load uses this energy to perform some useful function. But another way you can put it is that the load uses the energy from the source to do *work*. In doing the work, the load uses up the energy; this is why batteries "wear out" and have to be recharged or replaced. The amount of work done by a load depends on the amount of energy provided to the load and how fast the load can use the energy. In other words, with equal amounts of energy available, some loads can do more work in the same time than can others. Thus, some loads do work faster than others.

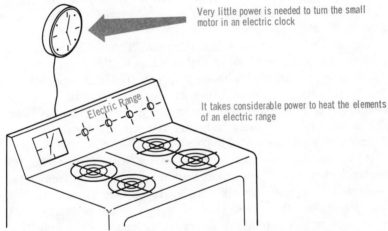

Very little power is needed to turn the small motor in an electric clock

It takes considerable power to heat the elements of an electric range

Power is the rate at which work is done. The faster work is done, the more power is used

The term *power* is used to describe how fast a load can do work. It can be defined as follows: *Power is the amount of work that can be done by a load in some standard amount of time, usually one second.* An important point you should remember is that the work done in an electric circuit can be *useful* work or it can be *wasted* work. In both cases, the rate at which the work is done is still measured in power. The turning of an electric motor is useful work, as is the heating of the element in an electric oven. The heating of the connecting wires or resistors in a circuit on the other hand, represents wasted work, since no useful function is performed by the heat. When power is used for wasted work, it is said to be *dissipated*.

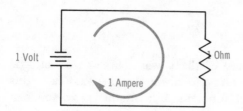

1 watt of power is used when 1 ampere flows through a potential difference of 1 volt. Therefore, in the circuit, 1 watt of power is used

the unit of power

Since power is the rate at which work is done, it has to be expressed in the units of *work* and *time*. You undoubtedly know that the basic unit of time is the second. However, you might not know the unit of work. For our purposes here, the unit of work will only be defined. You can find a description of how it is derived in many elementary physics books.

The unit of electrical work is the *joule*. This is the amount of work done by one *coulomb* flowing through a potential difference of 1 volt. Thus, if 5 coulombs flow through a potential difference of 1 volt, 5 joules of work are done. You can see that the time it takes these coulombs to flow through the potential difference has no bearing on the amount of work done. Whether it takes one second or one year, 5 joules of work are still done.

It is more convenient when working with circuits to think of amperes of current rather than coulombs; and as you remember from Volume 1, 1 ampere equals 1 coulomb passing a point in 1 second. Using amperes, then, one joule of work is done in one second when 1 ampere moves through a potential difference of 1 volt. This rate of 1 joule of work in 1 second is the basic unit of *power*, and is called a watt. Therefore, *a watt is the power used when one ampere of current flows through a potential difference of one volt.*

Mechanical power is usually measured in units of *horsepower*, abbreviated *hp*. You will sometimes find it necessary to convert from watts to horsepower and vice versa. To convert from horsepower to watts, multiply the number of horsepower by 746. And to convert from watts to horsepower divide the number of watts by 746.

CONVERSION OF UNITS

1000 watts (w)	= 1 kilowatt (kw)
1,000,000 watts (w)	= 1 megawatt (Megaw)
1000 kilowatts (kw)	= 1 megawatt (Megaw)
1 watt (w)	= 0.00134 horsepower (hp)
1 horsepower (hp)	= 746 watts (w)

equations

From the definition of the watt, you know that 1 watt of power is used when 1 ampere flows through a difference of 1 volt. Then, if 2 amperes flow through a difference of 1 volt, 2 watts of power must be used. Or, if 1 ampere flows through a difference of 2 volts, 2 watts must be used. In other words, the number of watts used is equal to the number of amperes of current times the potential difference. This is expressed in equation form as:

$$P = EI$$

where P is the power used, in watts; E is the potential difference, in volts; and I is the current, in amperes.

The equation is sometimes called Ohm's Law for power, because it is similar to Ohm's Law. This equation allows you to find the power used in a circuit or load when you know the values of current and voltage. Two other useful forms of the equation are:

$$E = P/I$$

which is used when you know the power and current and want to find the voltage; and:

$$I = P/E$$

which is used to find the current when you know the power and voltage. You can see, then, that with these three equations you can calculate the power, voltage, or current in a circuit as long as you know the value of the other two.

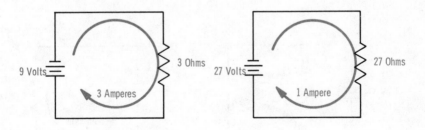

Since the equation for power is $P = EI$, both of these circuits use the same power:

$$P = EI = 9 \text{ volts} \times 3 \text{ amperes} = 27 \text{ watts}$$

$$P = EI = 27 \text{ volts} \times 1 \text{ ampere} = 27 \text{ watts}$$

resistance equations

There will be times when you have to find the power in a circuit and all you know are the voltage and resistance. You could first apply Ohm's Law to find the circuit current, but this takes time. It is easier to use an equation that gives power in terms of *voltage* and *resistance*. Since the power equations and Ohm's Law are similar, an equation can be easily derived.

You know that $P = EI$ and that $I = E/R$. So if you replace the I in the power equation with its Ohm's Law equivalent, E/R, you get:

$$P = E \times (E/R) = E^2/R$$

With this equation, all you have to know are the resistance and voltage and you can calculate the power. The term E^2 is pronounced "E squared," and means E multiplied by itself.

In the same way that the equation $P = E^2/R$ was derived, an equation can be obtained giving power in terms of *current* and *resistance*. Such an equation would be used when you know the current and resistance and are asked to find the power. To derive this equation, you use $E = IR$. When these are combined, you get:

$$P = IR \times I = I^2R$$

Two steps are needed to find the power used in the first circuit below using the equations on page 2-45. How would you do it now in only one step? Since power is the unknown and voltage and resistance are the knowns, you have to use the equation that relates power to voltage and resistance. The equation is

$$P = E^2/R = (100)^2/10 = (100 \times 100)/10 = 1000 \text{ watts}$$

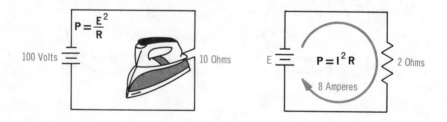

What is the power consumed in the second circuit? Now you would use the equation $P = I^2R$, since the current and resistance are known:

$$P = I^2R = 8 \times 8 \times 2 = 128 \text{ watts}$$

power losses

The power consumed in a circuit indicates how much work is done in that circuit. But you will remember that this work is not always useful. Much of it may be wasted, or lost. Power used to perform wasted work is, therefore, lost or *dissipated* power. In terms of the power source, lost power represents electrical energy that is not being used productively. And as you know, the production of electrical energy, whether by a battery or an electrical generator, costs money. It is important, therefore, that power losses in any electric circuit be kept as small as possible.

The most common loss of power in an electric circuit is the heat produced when current flows through a resistance. The exact relationship between the three quantities of heat, current, and resistance is given by the power equation:

$$P = I^2R$$

where P is the rate at which the heat is produced. You can see from the equation that you can decrease the amount of heat produced by lowering either the current or the resistance.

This I^2R heating, as it is often called, takes place in the circuit wires as well as in resistors. It is normally very small in the wires, since material and wire sizes are used that have low resistance values. In a resistor, little can be done about the I^2R heating inasmuch as the circuit current and the resistance value of the resistor cannot usually be changed without affecting the operation of the circuit.

In some electrical appliances, such as toasters and irons, the I^2R heating is needed and so does not represent a power loss.

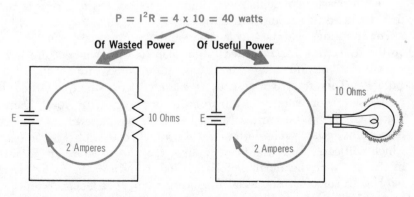

$P = I^2R = 4 \times 10 = 40$ watts

Of Wasted Power   Of Useful Power

E   10 Ohms   E   10 Ohms

2 Amperes   2 Amperes

I^2R heating usually represents wasted power. Sometimes, though it does useful work

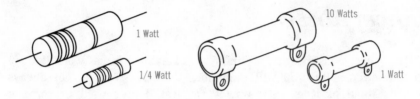

Often, power ratings are not marked on resistors, but indicated by their physical size. However, the sizes used for particular wattages vary not only between different types of resistors but also between manufacturers, so it might be difficult to judge the power rating. Check the manufacturer's list of characteristics

power rating of resistors

If too much current flows through a resistor, the heat caused by the current will damage or destroy the resistor. This heat is caused by I^2R heating, which you know is power loss expressed in watts. Therefore, every resistor is given a *wattage*, or *power*, rating to show how much I^2R heating it can take before it burns out. This means that a resistor with a power rating of 1 watt will burn out if it is used in a circuit where the current causes it to dissipate heat at a rate greater than 1 watt.

If you know the power rating of a resistor and want to find the maximum current it can carry, you can use an equation derived from $P = I^2R$:

$$P = I^2R \text{ becomes } I^2 = P/R, \text{ which becomes } I = \sqrt{P/R}$$

Using this equation, you can find the maximum current that can be carried by a 1-ohm resistor with a power rating of 4 watts:

$$I = \sqrt{P/R} = \sqrt{4/1} = \sqrt{4} = 2 \text{ amperes}$$

If such a resistor conducts more than 2 amperes, it will dissipate more than its rated power and will burn out.

Power ratings assigned by resistor manufacturers are usually based on the resistors being mounted in an open location where there is free air circulation, and where the temperature is not higher than 40°C (104°F). Therefore, if a resistor is mounted in a small, crowded, enclosed space, or where the temperature is higher than 40°C, there is a good chance it will burn out even before its power rating is exceeded. Also, some resistors are designed to be attached to a chassis or frame, which will carry away the heat. If these types of resistors are mounted in the air, their heat will not be carried away and they will become too hot. In some cases, special mounting clamps, called heat sinks and cooling fins, are used to carry the heat away. Sometimes fan blowers are used to cool the resistors to increase their power rating.

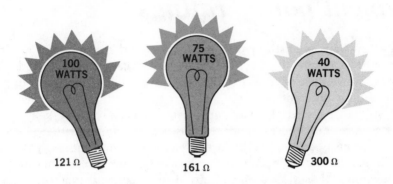

The wattage, or power, rating of a lamp tells how bright the lamp will be when used in a circuit. Actually, it is a measure of the I^2R heating of the lamp filament, which depends on the resistance of the filament

incandescent lamp power rating

From Volume 1, you know that an incandescent lamp is made by enclosing a resistance element, called a filament, in a glass bulb. When the lamp is connected into a circuit, current flows through the filament, and I^2R heating takes place. The heat is so severe that the filament becomes white-hot, and gives off light. The *more* the filament is *heated,* the *more* light the lamp gives off. You can see then that a convenient way of rating electric lamps is according to the I^2R heating they produce. This is exactly what is done by commercial lamp manufacturers. They stamp on each lamp the value of I^2R heating, in watts, that will be produced when the lamp is connected to a standard source of power. And, then, when you buy a lamp according to its *wattage rating,* you are really selecting it according to its light output.

You may wonder what the physical difference is between a 40-watt lamp and a 100-watt lamp. You should see from the equation $P = EI$ that the higher wattage bulb either has more current flowing through it or is connected to a higher source of voltage, or maybe both. However, you know that in most cases, such as in your own home, the source voltage is set by the local power company and cannot be changed. This means that the 100-watt lamp must conduct more current. To do this, it must have a lower resistance than the 40-watt lamp. Thus, the wattage and, therefore, the light output of an electric lamp depends on the resistance of the lamp filament. The *higher* the *resistance,* the *lower* the rated *wattage;* and the *lower* the *resistance,* the *higher* the rated *wattage.*

typical power ratings

As you have seen, the power ratings used for resistors and electric lamps are measures of the I^2R heating that takes place. Although the power rating always measures I^2R heating, its *practical* meaning is different for different devices.

Many other electrical devices are selected on the basis of their power ratings, especially those appliances that use heat to do their job. These include irons, toasters, heaters, ovens, etc. For most of these appliances, the larger the power rating, the more heat is produced. This means, for example, that a 1500-watt electric heater puts out more heat than a 1000-watt heater, and can therefore heat a larger area. Appliances with the highest power ratings, however, are not always the best. Toasters with power ratings of 10,000 watts or more could be made, but they would not toast your bread, they would burn it to a crisp almost instantly.

The power rating of working equipment such as motors, though, are not based on I^2R losses. They are based mostly on the power they can use to do mechanical work. A *1-horsepower* motor uses *746 watts* of electrical power, plus whatever power is dissipated because of I^2R losses. A ¼-hp motor needs at least 186.5 watts of electrical power.

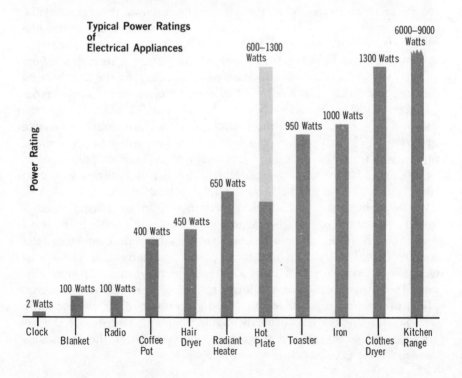

Typical Power Ratings of Electrical Appliances

the kilowatt-hour

Practically all of the electricity used in this country is produced, or generated, by large electric companies. From the power stations where it is generated, the electricity is distributed to the users by a complicated arrangement of wires, cables, and other devices. This distribution system ends at the individual factories, stores, or homes where the electricity is to be used. Since the electric companies are in the business of selling electricity, they must have some way of knowing *how much* electricity is used by each of their customers. Otherwise, they would have no way of knowing how much money to charge. They do this by supplying a meter to every user. The meter is usually located where the electricity enters a home, apartment, or factory, and measures the electricity used.

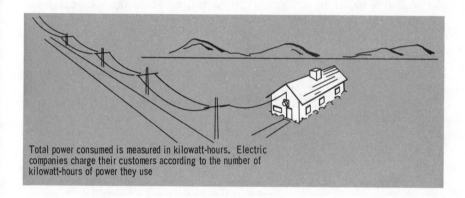

Total power consumed is measured in kilowatt-hours. Electric companies charge their customers according to the number of kilowatt-hours of power they use

Electricity itself cannot be measured, since as you know, it is really only a phenomenon. Current and voltage can be measured, but as you will learn later, to charge a user on the basis of current or voltage alone is impractical. Instead, every customer is charged on the basis of how much *work* is done by the electricity he uses.

You remember that the rate at which work is done is measured in watts. So to find the *total work* done, which is the same as the total power consumed, you multiply the rate of doing work (watts) by the total time during which the work is done. Thus, if a 100-watt lamp burns for one hour, the total work is 100 watts times 1 hour, or 100 watt-hours. This, then, is the way the electric companies measure and charge for electricity.

The watt-hour is a small unit. If it was used to indicate total power consumed, large numbers would result. So instead, units of *kilowatt-hours* are used. Each kilowatt-hour is equal to 1000 watt-hours.

summary

☐ Power describes how fast a load can do work. It is the amount of work that can be done in some standard amount of time. ☐ The unit of electrical work is the joule, which is the amount of work done by 1 coulomb flowing through a potential difference of 1 volt. The unit of time used is the second. ☐ One basic electrical power unit is the rate of 1 joule of energy consumed per second. It is called the watt. The watt can also be expressed as the power used when one ampere of current flows through a potential difference of 1 volt. ☐ The unit for mechanical power is the horsepower, hp. ☐ One horsepower equals 746 watts.

☐ Power can be calculated in terms of voltage, current, and resistance by three equations: $P = EI$, $P = I^2R$, and $P = E^2/R$. ☐ The three equations can be rearranged to solve for voltage, current, and resistance: $E = P/I$ and $E = \sqrt{PR}$; $I = P/E$ and $I = \sqrt{P/R}$; and $R = E^2/P$ and $R = P/I^2$. ☐ Work done in an electric circuit can be useful or wasted work. When power is used for wasted work, it is said to be dissipated. ☐ Power losses (dissipated power) are usually in the form of heat, called I^2R heating.

☐ The power, or wattage, ratings of resistors are usually at 104°F, and in open air. These conditions must be noted when mounting a resistor. ☐ Wattage ratings of incandescent lamps are determined by the resistance of the filament. The higher the wattage rating, the lower the resistance. ☐ The amount of light given off by an incandescent lamp depends on its wattage rating. The higher the wattage rating, the more light it yields. ☐ Electrical work or energy is used by electric companies to measure and charge for electricity. The basic unit is the watt-hour, but the kilowatt-hour (1000 times larger) is more convenient.

review questions

1. Define the following: *joule, watt, watt-hour,* and *kilowatt-hour.*
2. How many joules are there in 1 kilowatt-hour?
3. Draw a "Pie" diagram for P, E, and I.
4. A certain motor consumes 1492 watts. What is the power rating in hp? In kw?
5. If it was desired to obtain more light in a room, should a bulb with a smaller or greater wattage rating be used? Why?
6. What is meant by I^2R *losses*?
7. What is the equation for finding P, if E and I are given?
8. What is the equation for finding E, if P and R are given?
9. What is the equation for finding I, if R and P are given?
10. What is the equation for finding R, if P and E are given?

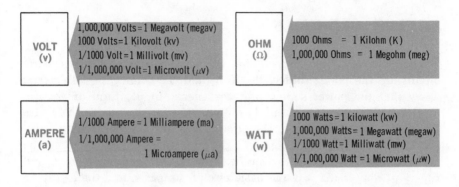

Often, the basic units of the volt, ohm, ampere, and watt are either too large or too small for practical use. Fractional or multiple values of the units are used instead:

mega = 1,000,000; kilo = 1000; milli = 1/1000; and micro = 1/1,000,000

basic electrical units

Based on what you have learned up to now, you can see that in every d-c circuit there are four electrical quantities that you will work with most often. These are (1) the source *voltage* applied to the circuit, (2) the *resistance* present in the circuit, (3) the *current* that flows in the circuit, and (4) the *power* consumed by the circuit. The units in which each of these quantities are normally expressed make up what is called the practical system of units. These units are:

Quantity	Unit
Voltage	Volt
Resistance	Ohm
Current	Ampere
Power	Watt

The *volt* is defined in terms of electrical work as follows: When an emf moves 1 coulomb of electrons (6.28 billion billion electrons) to do 1 joule of work, the emf has a potential difference of 1 volt.

The *ampere* is defined in terms of coulombs of charge. A current of 1 ampere flows when 1 coulomb passes a given point in 1 second. You can see that the ampere is a measure of the rate of flow.

The *ohm* is defined in terms of the volt and the ampere. A material has a resistance of 1 ohm when an emf of 1 volt causes a current of 1 ampere to flow through it.

The *watt* is also defined in terms of the volt and the ampere. It is the power used when 1 ampere of current flows through a potential difference of 1 volt.

series circuits

Thus far, you have learned that the basic elements of the electric circuit are the power source, the load, and the connecting wires. You know that the power source provides energy in the form of voltage, and that this voltage causes current to flow in the circuit. You also know that the resistance in the circuit, whether it is the resistance of the load or of the wires, opposes the current flow. In addition, you learned how the load uses electrical energy to perform some useful function. This is the way the basic circuit, its parts, and their purpose were presented. Once you understood the basic circuit, the relationships that exist between voltage, current, resistance, and power were explained.

Throughout the presentation of this material, all the circuits described had one basic thing in common. That is, there was *one* and only one *path* for the current to follow as it flowed from the negative terminal of the power source, through the circuit, and back to the positive terminal of the source. There was *never* any point in a circuit where the current could divide and take more than one path. As a result, exactly the same current flowed through every part or device in the circuit.

This type of one-path circuit is called a *series circuit.*

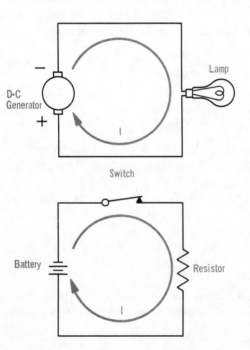

In a series circuit, the same current flows through every part. It makes no difference how many parts or devices there are. As long as the identical current passes through each, it is a series circuit

series loads

Up until now, you have studied circuits that have a single load, such as one resistor or one lamp. The resistance of this single load has been the total circuit resistance, and the power consumed in the circuit was the power used by this one load. In actual practice, however, you will very often find that a circuit has more than one load. It may have two resistors, or a resistor and a lamp, or maybe even five resistors and two lamps. In fact, there is almost no limit to the number of individual loads that a circuit may have.

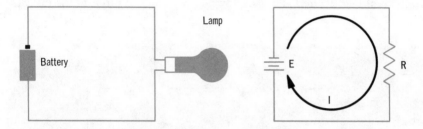

Previously, we have examined circuits that have a single load. The total circuit resistance was the resistance of the load, and the total circuit power was the power used by the load

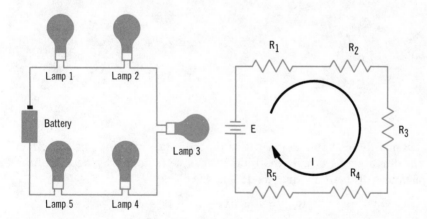

More than one load can be used in a circuit. If the total current (I) flows through each of them, they are series loads

If the loads are connected in the circuit in such a way that the total circuit current passes through each, then they are connected in *series*. They are *series loads*.

how series loads affect current

The current that flows in a circuit depends on the source voltage and the *total circuit resistance*. When there is only one load in a circuit, it usually provides the total circuit resistance. However, when series loads are used, the total circuit resistance is the *sum* of the resistances of each individual load. Thus, if a circuit has five loads connected in series, and each load is a 10-ohm resistor, the total circuit resistance is 5×10 or 50 ohms.

To find the current in a circuit that contains series loads, you first determine the total circuit resistance by adding the resistances of the individual loads. Then you use Ohm's Law $(I = E/R)$ to find the current.

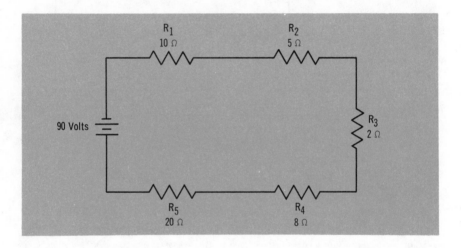

Since the current flow depends on the total resistance of the circuit, and for a series circuit the total resistance is found by adding the resistances of the individual loads, for this circuit:

$$R_{TOT} = R_1 + R_2 + R_3 + R_4 + R_5$$
$$= 10 + 5 + 2 + 8 + 20$$
$$= 45 \text{ ohms}$$

Once the total resistance is known, Ohm's Law can be used to find the circuit current:

$$I = E/R_{TOT}$$
$$= 90 \text{ volts}/45 \text{ ohms}$$
$$= 2 \text{ amperes}$$

series power sources

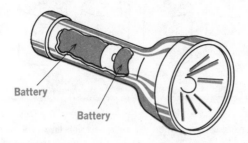

Frequently, the voltage required for operation of a device or circuit is larger than that of any available power source. In these cases, power sources can be used in series to give the needed voltage. An example of this is the common flashlight

You have probably at one time or another bought batteries for your car, your flashlight, your portable radio, or the flash unit for your camera. And as a result, you know that there are batteries with 1½-volt outputs, 6-volt outputs, 9-volt outputs, and 12-volt outputs, just to name a few. But you have probably never seen a 3-volt battery, a 15-volt battery, or a 29-volt battery. The reason is that it is unprofitable for battery manufacturers to mass produce batteries with every possible voltage output. So instead, they make certain standard sizes that can be *combined* to get most of the required voltage.

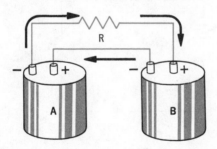

When two or more batteries are used in a circuit to produce a greater voltage than either battery can alone, then the batteries are connected in series. This produces *series power sources*.

When two batteries are connected in series, the negative terminal of one is connected to the positive terminal of the other. The other two terminals are connected to the circuit. As shown, current leaves the negative terminal of battery A, flows through the circuit, and enters the positive terminal of battery B. It then leaves the negative terminal of B and returns to the positive terminal of A.

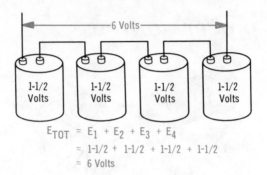

$$E_{TOT} = E_1 + E_2 + E_3 + E_4$$
$$= 1\text{-}1/2 + 1\text{-}1/2 + 1\text{-}1/2 + 1\text{-}1/2$$
$$= 6 \text{ Volts}$$

For power sources connected in series, the total voltage output is the sum of the individual source voltages

how series power sources affect current

When power sources are connected in series, the *total* voltage is equal to the *sum* of the *individual* source voltages. To find the current in a circuit containing power sources connected in series, you must therefore first find the total source voltage. Then you can use Ohm's Law $(I = E/R)$ to calculate the current.

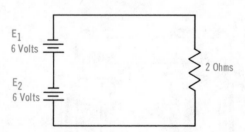

For the above circuit with series power sources, the total source voltage is

$$E_{TOT} = E_1 + E_2$$
$$= 6 + 6$$
$$= 12 \text{ volts}$$

To calculate the current,

$$I = E_{TOT}/R$$
$$= 12 \text{ volts}/2 \text{ ohms}$$
$$= 6 \text{ amperes}$$

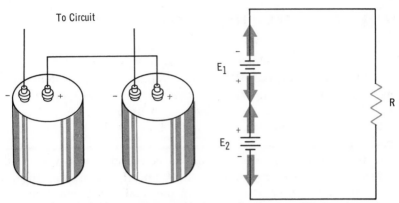

If the battery polarities are not connected in the same direction, they will be in series opposing, and will subtract from each other. Therefore,

$$E_{TOT} = E_1 - E_2$$

series-opposing power sources

If power sources are connected so that their polarities are not in the same direction, they will oppose each other. Power sources connected in this way are said to be *series-opposing*. The *total* voltage of series-opposing sources is the *difference* between the *individual* voltages:

$$E_{TOT} = E_{LARGER} - E_{SMALLER}$$

The polarity of the total source voltage will be the same as that of the larger battery, but if both batteries have the same voltage, then E_{TOT} will be zero, and no current will flow.

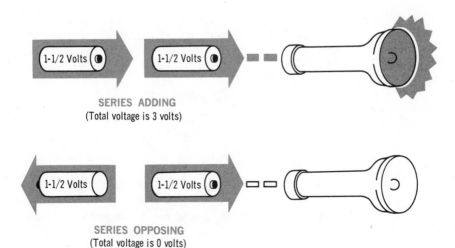

SERIES ADDING
(Total voltage is 3 volts)

SERIES OPPOSING
(Total voltage is 0 volts)

power consumption

Power is the rate at which a load does work. And when there is only one load in a circuit, the power used by that load is the total power consumed in the circuit. When a circuit has a number of loads connected in series, each of the individual loads uses power. Therefore, the total power consumed in the circuit is the sum of the power consumptions of each load.

The total power in a series circuit can be found in two ways. One is to calculate the power used by each load, and then add these together. The second, and easier way, is to find the total circuit resistance, and then calculate the power that is used by the total resistance.

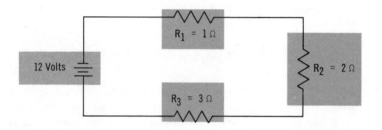

To find the power used by each of the loads in this circuit, you must first know the circuit current. But before you can find the current, you have to find the total circuit resistance.

Calculating Total Circuit Resistance:

$$R_{TOT} = R_1 + R_2 + R_3 = 1 + 2 + 3 = 6 \text{ ohms}$$

Calculating Circuit Current:

$$I = E/R_{TOT} = 12 \text{ volts}/6 \text{ ohms} = 2 \text{ amperes}$$

Calculating Power Used by Each Load:

$$P = I^2R$$
$$P_1 = 2 \times 2 \times 1 = 4 \text{ watts}$$
$$P_2 = 2 \times 2 \times 2 = 8 \text{ watts}$$
$$P_3 = 2 \times 2 \times 3 = 12 \text{ watts}$$

Calculating Total Circuit Power:

$$P_{TOT} = P_1 + P_2 + P_3 = 4 + 8 + 12 = 24 \text{ watts}$$

The total circuit power could also be found more simply by the equation:

$$P_{TOT} = EI = 12 \text{ volts} \times 2 \text{ amperes} = 24 \text{ watts}$$

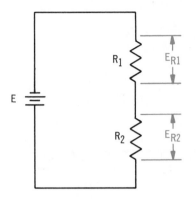

The total voltage supplied by a source is dropped across the circuit resistance. You can consider the voltage drop either as a loss of energy of the electron flow, or as the work done by the current when it flows through resistance. The energy lost is usually given off as heat. The total voltage drop equals the source voltage:

$$E_{SOURCE} = E_{R1} + E_{R2}$$

the voltage drop

You remember from Volume 1 that in a battery, a difference of potential is set up between the positive and negative terminals. This is done by chemically causing an *excess of electrons* at the *negative terminal* and a corresponding *lack of electrons,* or positive charges, at the *positive terminal.* When a wire or any conducting material is connected between the terminals, the difference of potential causes a force field, which we will call a force, to be sent down the wire at the speed of light. Electrons then flow from the negative terminal through the wire and back to the positive terminal under the pressure of the force. For every electron that leaves the negative terminal, another is produced chemically by the battery to take its place. Likewise, for every electron that arrives at the positive terminal and neutralizes a positive ion, another ion is produced by the battery to replace it. The battery thus keeps its difference of potential constant, although electrons are flowing.

Every electron at the negative battery terminal has been given energy by the battery. When the electron moves around the circuit it gives up the energy, so that when it arrives at the positive terminal it has lost all the energy the battery had given it. The electron loses its energy by giving it to the circuit resistance, usually in the form of heat.

Since the difference of potential across the battery terminals is normally given in volts, the energy lost by the electrons in the circuit resistance is also expressed in volts. Thus, if a resistor was connected across a 10-volt battery, 10 volts would be lost, or *dropped,* by the current flowing through the resistor. If two or more resistors were connected across the battery, some voltage would be dropped across each resistor, but the total voltage dropped would still be only 10 volts. Therefore, the total voltage drop in a circuit always equals the source voltage.

calculating voltage drops

In a series circuit, the *total* voltage dropped across all of the loads is equal to the source voltage. This is true whether there is one load or fifty loads. You can see, then, that for a fixed source voltage, the more loads there are, the less voltage will be dropped across each load.

Since the *voltage drop* across any load is the energy given to the load, the voltage that is dropped depends on the *current* flowing through the load and the *resistance* of the load. The greater the current, or the higher the resistance, the more voltage is dropped. And, the smaller the current, or the lower the resistance, the less voltage is dropped. This is shown by the equation $E = IR$. You will recognize this as one of the equations of Ohm's Law. It shows the relationships between the current, voltage, and the resistance of individual circuit components, as well as entire circuits.

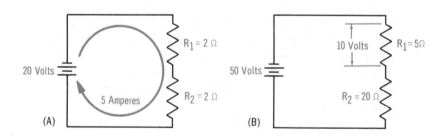

(A)

(B)

For circuit A the voltage drop across R_1 is

$$E_{R1} = IR_1 = 5 \text{ amperes} \times 2 \text{ ohms} = 10 \text{ volts}$$

Since you know that the drop across R_1 is 10 volts, and that the total drop must equal the source voltage, then the remaining voltage must be dropped across R_2. This can be found by:

$$E_{R2} = E_{TOT} - E_{R1} = 20 - 10 = 10 \text{ volts}$$

The voltage drops across R_1 and R_2, then, are the same: 10 volts each. But this makes sense since they both have the same resistance and carry the same current.

As shown in circuit B, sometimes you may know the resistance of a load and the voltage drop across it, and you are asked to find the current. You would do this by using the equation $I = E/R$. Therefore, the current through R_1 is

$$I = E_{R1}/R_1 = 10 \text{ volts}/5 \text{ ohms} = 2 \text{ amperes}$$

polarities

You remember that all voltages and currents have polarity as well as magnitude. In a series circuit there is only one current, and its polarity is from the negative battery terminal, through the circuit, to the positive battery terminal. The voltage drops across the loads also have polarities. The easiest way to find out what these polarities are, is to use the direction of the electron current as a basis. Then, where the electron current enters the load, the voltage is negative; and where it leaves the load, the voltage is positive. This holds true no matter how many loads there are in the circuit or what type they may be.

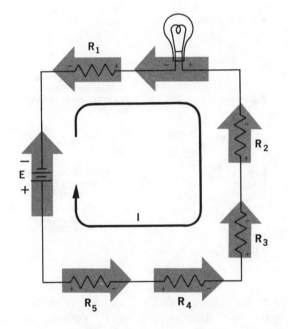

The voltage drops oppose the voltage from the source

Current enters the negative side of a load and leaves by the positive side. Current flow is from negative to positive within loads, and from positive to negative within power sources. This can be explained by the energy of the flowing charges. Inside of the source, their energy is increased, while inside the loads, their energy is decreased

The drop across the load, then, is opposite to that of the source. The voltage drops oppose the source voltage, and reduce it for the other loads. The reason for this is that each load uses energy, leaving less energy for the other loads.

polarity and voltage
of a point

Outside of a power source, current always flows from negative to positive. You may have wondered while studying the illustration on page 2-63 whether that was always true, inasmuch as the polarities seem to show that between the loads the current was flowing from positive to negative. Actually, each pair of polarities applies only to the load it is near. The plus and minus for one load has no relationship to the plus and minus of any other load. If you look at the illustration, this will probably be clear to you.

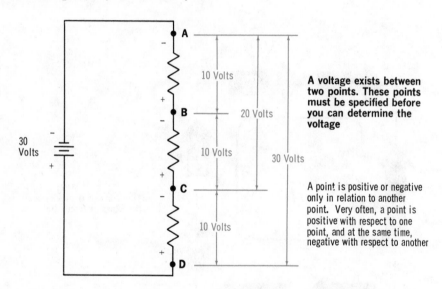

A voltage exists between two points. These points must be specified before you can determine the voltage

A point is positive or negative only in relation to another point. Very often, a point is positive with respect to one point, and at the same time, negative with respect to another

Point B has both a plus and a minus because it is positive with respect to point A, but negative with respect to point C. You can see, then, that to say a point is positive or a point is negative has no meaning. A point has to be positive or negative with respect to some other point.

In the same way, a voltage does not exist at a point. Voltage is a *difference of potential* between two points. This is shown in the illustration. If 10 volts is dropped across each resistor, the voltage between points A and B, B and C, and C and D are all 10 volts. But the voltage between A and C is the sum of the voltage drops between A and B, and B and C; or 20 volts. Similarly, the voltage between points A and D is 30 volts. You can see that to specify a voltage, you have to identify the *two points*.

the potentiometer circuit

You can see from the illustration on page 2-64 that in a circuit with series loads, different voltages exist between different points in the circuit. Later you will find that this is very useful, since it allows different values of voltage to be obtained from one source. Obviously, the more series loads there are, the more values of voltage there are in the circuit. But to get many voltages in this way requires a very large number of series loads.

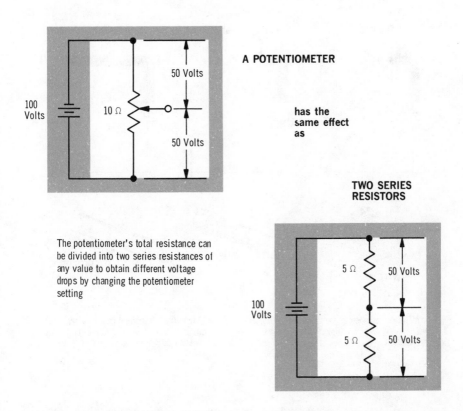

A POTENTIOMETER

has the same effect as

TWO SERIES RESISTORS

The potentiometer's total resistance can be divided into two series resistances of any value to obtain different voltage drops by changing the potentiometer setting

One way to get many voltages and still keep the number of loads to a minimum, is to use a potentiometer as the load. You remember from page 2-32 that a potentiometer has three terminals, and all three are connected into a circuit. The terminal that is connected to the movable arm of the potentiometer, therefore, actually divides the total resistance into two series resistances, and each resistance has its own voltage drop across it. By varying the resistance of the potentiometer, any value of resistance, and thus any voltage drop, can be obtained.

open circuits

A circuit has to provide a *complete* path from the negative terminal of the power source to the positive terminal in order for current to flow. A series circuit has only one path, and if it is broken, no current flows, and the circuit becomes an *open* circuit.

In a series circuit, all of the current flows through every point in the circuit. Therefore, if the circuit is open at ANY point, all current flow stops

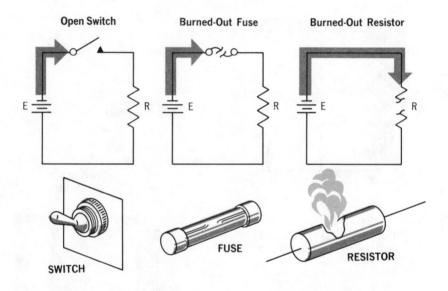

Circuits can be open by switches and fuses, as well as by damaged or broken parts or open connections

Since all current stops flowing when a series circuit opens, everything in a circuit that depends on the current is also affected

Circuits may be opened deliberately, such as by switches, or they may be opened as a result of some defect or trouble, such as a broken wire or a burned-out resistor. Since too much current in a circuit can damage the power source and the load, fuses are usually used in circuits to protect against damage. A fuse does its job by opening the circuit before high currents can do any harm.

Inasmuch as *no current flows* in an *open* series circuit, there are no voltage drops across the loads. There is also no power used by the loads, and so the total power consumed in the circuit is zero.

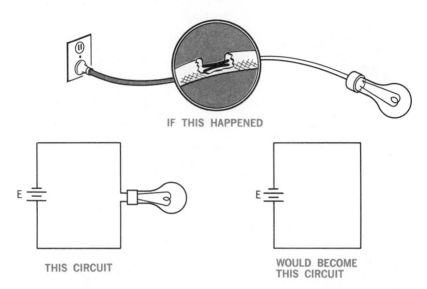

IF THIS HAPPENED

THIS CIRCUIT

WOULD BECOME
THIS CIRCUIT

A short circuit exists when current can flow from the negative terminal of the power source through connecting wires and back to the positive terminal of the power source without going through any load

short circuits

In a d-c circuit, the resistance is the only thing that opposes current flow. Therefore, if there was no resistance in a circuit, or if the resistance suddenly became zero, a very large current will flow. This condition of *no resistance* and very *high current* is called a *short circuit*.

From a practical standpoint, the resistance of a circuit cannot be reduced completely to zero. Even if a piece of silver wire with a large cross-sectional area was connected directly between the output terminals of a power source, there would still be some resistance in the circuit. This would consist of the resistance of the wire and the internal resistance of the power source. These resistances would be so low, though, that they would not limit the current flow very much. For example, if the combined resistance of the wire and the power source were 0.5 ohm and the source voltage was 100 volts, from Ohm's Law, the current would be: $I = E/R = 100$ volts$/0.5$ ohm $= 200$ amperes.

A short circuit is therefore said to exist whenever the resistance of a circuit becomes so low that the circuit current increases to the point where it can damage the circuit components. Current from short circuits can damage power sources, burn the insulation on wires, and start fires from the intense heat it produces in conductors. Fuses and other circuit breakers are the major means of protecting against the dangers of short circuits.

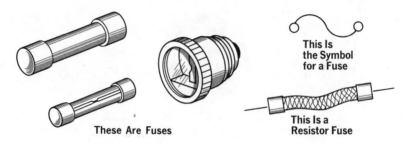

This Is the Symbol for a Fuse

These Are Fuses

This Is a Resistor Fuse

Fuses open circuits before the high currents caused by short circuits can do any damage. The maximum current a fuse can carry before it melts and opens the circuit is called the rating. Fuses are normally rated in amperes, such as a 1-ampere fuse, or a 5-ampere fuse. The rating is usually marked on the fuse. Resistor fuses also have a resistance rating.

the fuse

You know that fuses open circuits to prevent the high currents caused by short circuits from doing damage. A fuse, therefore, has to do three things: (1) it has to know, or sense, when a short circuit exists, (2) it has to open the circuit before any damage is done, and (3) it has to have no effect on the circuit during normal operation, that is, when no short circuit exists.

Basically, most fuses are pieces of soft metal conductors. or elements, contained in a housing. The fuse is connected into a circuit so that the fuse element is in series with the loads and power source. This means that the total circuit current flows through the fuse element. The element has very little resistance, and so has practically no effect on the circuit under normal conditions.

When a short circuit occurs, the current flowing through the fuse increases greatly; this causes the I^2R heating of the fuse element to increase. The fuse element has a low melting point, which means that it melts at a lower temperature than ordinary wire conductors. And when the heat caused by the short-circuit current reaches the melting point of the fuse element, the element melts and opens the circuit. The current that a fuse can carry before it melts depends on the material used for the element and its cross-sectional area. When a fuse melts, it is no longer any good, and must be replaced.

Sometimes *resistors* are made in such a way that they also act as *fuses*. A wire is used in the resistor which melts when it gets too hot. This happens when enough current flows to exceed the power rating of the resistor. In other cases, a special *fuse wire* is used to make the connection in the circuit that has to be protected. It is a plain length of insulated, thin wire that opens when it gets too hot.

circuit breakers

The trouble with fuses of any kind is that they must burn out to accomplish their purpose. This means they must be replaced, and must be stocked nearby so that you will have them for replacement when they are needed. Circuit breakers are another kind of protective device. They have the same function as fuses, but work in a different way so that they do not have to go bad to work. The *electromagnetic circuit breaker* uses the principle of magnetism to open a circuit when too much current flows. A set of switch contacts inside the circuit breakers is normally kept closed by an armature. A spring tries to pull one contact away, but the armature holds it in place. When too much current flows through the coil, a magnetic field builds up and attracts the armature to the coil. This releases the contact, which is pulled away to open the circuit. The circuit breakers are made with a toggle lever to reset the contacts and armature so that the unit can still be used.

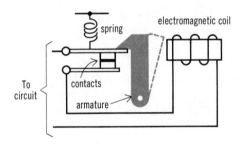

Electromagnetic Circuit Breaker
When too much current flows in the circuit breaker the magnetic field of the coil pulls in the armature. This allows the spring to pull the contacts apart to open the circuit

A *thermal circuit breaker* uses heat from the high current to work. One of the contact arms in this type of breaker is made of a temperature-sensitive metal that bends when it gets too hot. When it bends far enough, it will release the contact pulled by the spring. This kind of circuit breaker can also be reset after the bent metal cools and straightens.

Thermal Circuit Breaker
The thermal contact bends from the heat when too much current goes through it. This allows the other contact to be pulled **away** by the spring. Some thermal breakers do not use a spring to pull away the other contact. With these, when the thermal contact cools, it moves back in position to close the circuit again automatically

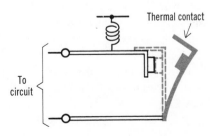

solved problems

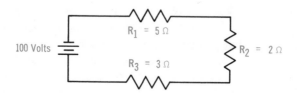

Problem 1. *In the circuit, what is the current and the total power consumed?*

Since in a series circuit, $I = E/R_{TOT}$, you have to find the total resistance before you can calculate the circuit current.

$$R_{TOT} = R_1 + R_2 + R_3 = 5 + 2 + 3 = 10 \text{ ohms}$$

With the total resistance, you find the current by Ohm's Law:

$$I = E/R_{TOT} = 100 \text{ volts}/10 \text{ ohms} = 10 \text{ amperes}$$

You can find the power consumed in the circuit in a variety of ways. One is by calculating the power used by each resistor, and then adding them. Another way is by directly calculating the power used by the total circuit resistance of 10 ohms. And still another way is by just using the source voltage and circuit current. No matter which method you use, the first step is to decide which equations for calculating power you should use. You remember that these equations are

$$P = EI \qquad P = I^2R \qquad P = E^2/R$$

If you decide to find the total power by first finding the power used by each individual resistor, the equation $P = I^2R$ should be used. The reason for this is that you already know the current through each resistor and the resistor value. Thus,

$$P_{TOT} = I^2R_1 + I^2R_2 + I^2R_3$$
$$= (100 \times 5) + (100 \times 2) + (100 \times 3) = 1000 \text{ watts}$$

To find the total power using the total circuit resistance of 10 ohms, you would also use the equation $P = I^2R$, since the current I and total resistance R_{TOT} are known.

$$P_{TOT} = I^2R_{TOT} = 100 \times 10 = 1000 \text{ watts}$$

Since both the circuit current and the source voltage are known, the power consumed in the circuit can also be found using the equation $P = EI$.

$$P = EI = 100 \text{ volts} \times 10 \text{ amperes} = 1000 \text{ watts}$$

solved problems (cont.)

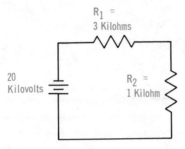

Problem 2. *In this circuit, what is the current?*

As you can see, the values of the source voltage and the resistances are given in multiples of the basic units. Whenever this happens, always convert to the basic units before trying to solve the problem.

You will remember from page 2-53 that 1 kilovolt equals 1000 volts. The source voltage in the circuit is therefore 20×1000, or 20,000 volts. Also you should recall that 1 kilohm equals 1000 ohms. Thus, the value of resistor R_1 is 3×1000, or 3000 ohms; and the value of resistor R_2 is 1×1000, or 1000 ohms.

Once you have converted the given values into the basic units, Ohm's Law is used to find the current.

$$I = E/R_{TOT} = E/(R_1 + R_2)$$
$$= 20{,}000 \text{ volts}/(3000 \text{ ohms} + 1000 \text{ ohms}) = 5 \text{ amperes}$$

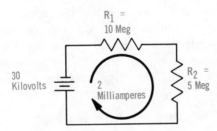

Problem 3. *Convert the quantities given on this circuit to the basic units.*

You know that 1 kilovolt equals 1000 volts. Therefore, 30 kilovolts is 30×1000, or 30,000 volts. One Meg equals 1,000,000 ohms. Therefore, 10 Meg is $10 \times 1{,}000{,}000$, or 10,000,000 ohms. And 5 Meg is $5 \times 1{,}000{,}000$, or 5,000,000 ohms. One milliampere equals 1/1000 ampere. Therefore, 2 milliamperes equals $2 \times 1/1000$, or 2/1000 (0.002) ampere.

solved problems (cont.)

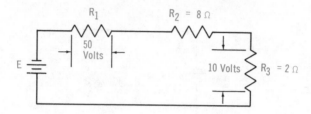

Problem 4.   What is the source voltage in this circuit?

You know that from Ohm's Law the equation for voltage is $E = IR$. And so, after examining the circuit diagram, you can see that one way to find the source voltage is to first find the circuit current and then the resistance value of R_1. You could then use $E = IR_{TOT}$ to calculate the source voltage. This would require the following steps:

1. Using the equation $I = E_3/R_3$ to find the current through R_3, which is also the total circuit current, since this is a series circuit.

   $$I = E_3/R_3 = 10 \text{ volts}/2 \text{ ohms} = 5 \text{ amperes}$$

2. Using the equation $R_1 = E_1/I$ to find the resistance of R_1.

   $$R_1 = E_1/I = 50 \text{ volts}/5 \text{ amperes} = 10 \text{ ohms}$$

3. Using the equation $E = IR_{TOT}$ to calculate the source voltage.

   $$E_{TOT} = IR_{TOT} = I(R_1 + R_2 + R_3) = 5(10 + 8 + 2) = 100 \text{ volts}$$

Another slightly easier way to solve this problem is to first find the circuit current, and then calculate the voltage drop across R_2. You would then know the voltage drop across each individual load in the circuit. And you remember that one of the basic laws of series circuits is that the sum of the individual voltages drops is equal to the applied source voltage.

Thus, to find the circuit current, you apply the equation $I = E/R$ to resistor R_3.

$$I = E_3/R_3 = 10 \text{ volts}/2 \text{ ohms} = 5 \text{ amperes}$$

Since it is a series circuit, the same current flows through resistor R_2. To find the voltage drop across R_2, then, you use the equation $E_2 = IR_2$:

$$E_2 = IR_2 = 5 \text{ amperes} \times 8 \text{ ohms} = 40 \text{ volts}$$

Now that you know the voltage drops across R_1, R_2, and R_3, the source voltage is simply the sum of the three voltage drops:

$$E_{SOURCE} = E_{R1} + E_{R2} + E_{R3} = 50 + 40 + 10 = 100 \text{ volts}$$

solved problems (cont.)

Problem 5. In the circuit, how much power is consumed by resistor R when the switch is opened?

Obviously, no power is used, since for power to be used, current must flow; and when the switch is opened, there is no current flow anywhere in the circuit.

Problem 6. If 5 amperes is the most current that the fuse can carry before it melts and opens the circuit, what is the lowest value at which R can be set?

The lowest value of R is the value which allows 5 amperes to flow, or

$$R = E/I = 200 \text{ volts}/5 \text{ amperes} = 40 \text{ ohms}$$

For any lower value of resistance, the current will be greater than 5 amperes and the fuse will "blow."

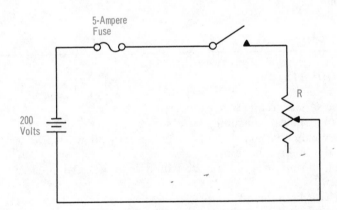

Problem 7. If the fuse does "blow" and opens the circuit, what will be the voltage drop across R?

No voltage will be dropped across R after the fuse blows, since there will be no current flow in the circuit.

Problem 8. If R is a 1000-watt resistor, and it is set for 20 ohms, there will be 10 amperes of current flow. This will cause R to dissipate 2000 watts of power. Will R burn out?

No. Current in the circuit will actually not reach 10 amperes since the fuse will blow at 5 amperes and open the circuit.

summary

□ A series circuit has only one path for the current. □ If a series circuit is broken, the circuit is open, and no current flows. □ Series loads are connected so that the total current passes through each. □ The total circuit resistance for series loads is the sum of the individual resistances.

□ Series power sources consist of two or more batteries connected to produce a greater voltage than either battery can alone. □ The total voltage for series power sources is the sum of the individual power sources. □ For power sources in series-aiding, the battery polarities are in the same direction, and the voltages add. □ In series-opposing, the battery polarities are in opposite directions, and the voltages subtract.

□ The total power consumed in a series circuit is the sum of the powers used by the individual loads. □ The total voltage dropped across all of the loads in a series circuit is equal to the source voltage. □ The polarity of a voltage drop across a load is opposite to that of the source. □ Voltage-drop polarity is determined by the current direction. The end of a load where current enters is negative; where it leaves positive. □ Polarity is always with respect to a reference point.

review questions

1. Define *series circuit*. *Series load*.
2. The total power consumed in a series circuit is 100 watts. If there are two loads, and one consumes 35 watts, what does the other consume?
3. In the circuit of Question 2, which load has the greater current?
4. With two batteries of 9 volts, and 1.5 volts, show how you can obtain supply voltages of 10.5, 7.5, −10.5, and −1.5 volts.

For Questions 5 to 10, consider a circuit consisting of five resistors, whose values are 5, 4, 3, 2, and 1 ohms, in series with a 30-volt battery.

5. What is the total resistance of the circuit?
6. What is the circuit current?
7. What is the voltage drop across the 3-ohm resistor?
8. What is the total power consumed in the circuit?
9. What is the power consumed by the 4-ohm resistor?
10. Answer Questions 6 to 9 with the circuit open.

parallel circuits

All of the circuits you have studied so far have been series circuits. The current flowing at every point in these circuits was the same. Thus, once you found the current at any point, you knew what the current was everywhere else in the circuit. It would greatly simplify the analysis of d-c circuits if all circuits had this feature, but this is not the case.

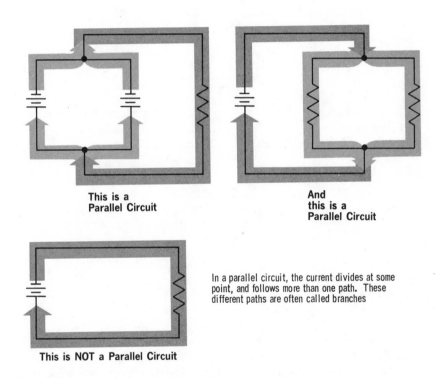

**This is a
Parallel Circuit**

**And
this is a
Parallel Circuit**

This is NOT a Parallel Circuit

In a parallel circuit, the current divides at some point, and follows more than one path. These different paths are often called branches

There is a large number of circuits in which the current is not the same at every point. In these circuits, there can be any number of *different currents*. All of the currents have the same polarity, since they are d-c circuits, but their magnitudes can vary greatly. These circuits are called *parallel circuits*, and can be defined as follows: *A parallel circuit is one in which there are one or more points where the current divides and follows different paths.*

When circuit components are connected in such a way that they provide different current paths, the components are said to be *connected in parallel.*

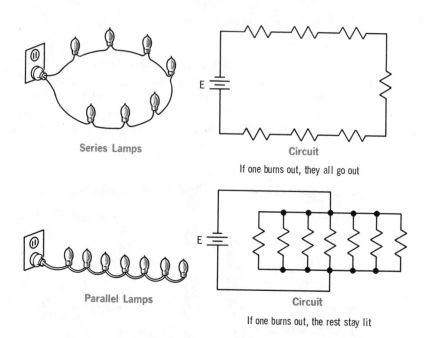

Series Lamps

Circuit

If one burns out, they all go out

Parallel Lamps

Circuit

If one burns out, the rest stay lit

If you have Christmas tree lamps, you are familiar with one of the basic features of parallel loads. That is, unlike series loads, if one load or branch becomes open, current still flows through other loads

parallel loads

In a series circuit, every load has the same current flowing through it; and this is the same current that both leaves and enters the power source. Very often, however, loads are connected in a circuit so that the *current* from the power source is *divided* between the loads, with only a *portion* of the current flowing through each load. The loads are then said to be connected in parallel, and the circuit is a parallel circuit.

In the parallel circuit, each load provides a separate path for current flow. The separate paths are called branches, and the current flowing in each branch is called *branch current*. Since the current that leaves the power source is divided between the branches, it is obvious that the current in any branch is less than the power source current.

If one branch of a parallel circuit is broken, or opened, current still flows in the circuit since there are still one or more complete paths for current flow through the other branches.

voltage drop across parallel loads

You will remember from series circuits that a portion of the source voltage is dropped across each series load. And the sum of these individual voltage drops is equal to the source voltage. When loads are connected in parallel, voltage is also dropped across each load. Instead of a portion of the source voltage being dropped across each load as in a series circuit, however, the *entire source voltage* is dropped across each. The reason for this is that all parallel loads are connected together directly across the power source.

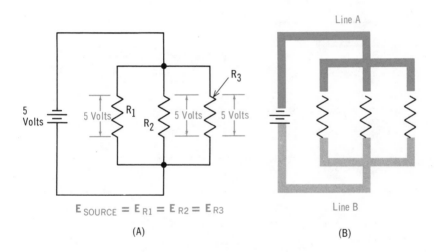

$$E_{SOURCE} = E_{R1} = E_{R2} = E_{R3}$$

(A)

(B)

Examining the circuit diagram (A) above, you can see that the entire source voltage is dropped across each branch, if you consider the lines that represent the connecting wires as lines of equal potential. In other words, the lines that connect the load and the power source have the same potential all along their length. This is shown on diagram (B).

Line A is at the potential of the negative battery terminal all along its length. Line B is at the potential of the positive battery terminal all along its length. The difference of potential between the battery terminals is 5 volts, so potential difference across each load must also be 5 volts.

effect of parallel loads on current

In a parallel circuit, the current leaves the power source, divides at some point to flow through the branches, and then recombines and flows back into the power source. There are thus *two* types of current in a parallel circuit: the current that leaves and reenters the power source, called the *total current;* and the *branch currents*. Since the total current divides to become the branch currents, the *sum* of the branch currents must be equal to the total current.

$$I_{TOT} = I_1 + I_2 + I_3$$

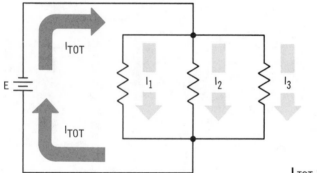

$$I_{TOT} = I_1 + I_2 + I_3$$

The total current in a parallel circuit is equal to the sum of the branch currents

Branch currents are determined by the resistance of the branch and the voltage across it. Since all branches have the same voltage across them, the more resistance there is in a branch, the less current there will be. And likewise, the less resistance there is, the greater the current will be. The total current in a parallel circuit depends on the source voltage and the total resistance of the circuit. As you will soon see, adding parallel loads to a circuit decreases the total circuit resistance, and thus increases the total current.

calculating branch currents

Each branch in a parallel circuit carries a separate current. Inside of a branch, though, the current is the same at every point. Every branch, therefore, has a voltage across it that is equal to the source voltage, a resistance, and a current that is the same at every point. You will recognize this as being similar to a series circuit. And as a matter of fact, this is exactly how you calculate the currents in the branches of parallel circuits. You take *one* branch at a time, consider it a series circuit, and use the equations for Ohm's Law.

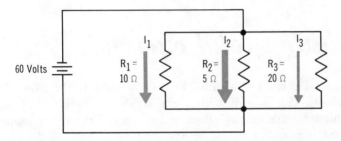

To calculate the branch currents of a parallel circuit, consider each branch as a separate series circuit, and then apply Ohm's Law. Thus, you would handle the above circuit as three separate circuits

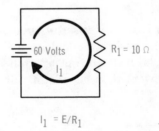

$$I_1 = E/R_1$$

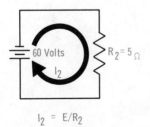

$$I_2 = E/R_2$$

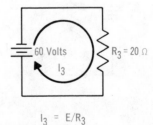

$$I_3 = E/R_3$$

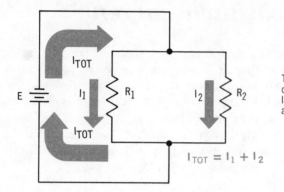

The total current in a parallel circuit can be found by calculating all of the branch currents, and adding them . . .

$$I_{TOT} = I_1 + I_2$$

calculating total current

Since the total current in a parallel circuit is equal to the sum of the branch currents, one way of finding the total current is to calculate all of the branch currents and then add their values. As an example of this, in the illustration on page 2-79, the total current in the circuit is the sum of the currents in the three branches, or 6 amperes plus 12 amperes plus 3 amperes, for a total of 21 amperes. Often it is easier to find the total current in a parallel circuit by calculating the total resistance in the circuit and then using the equation for Ohm's Law:

$$I_{TOT} = E/R_{TOT}$$

Based on what you know about series circuits, you may think that this would always be the easiest method to use. This is not true, though, since finding the total resistance of a parallel circuit is sometimes just as much, or more, work than calculating all of the branch currents.

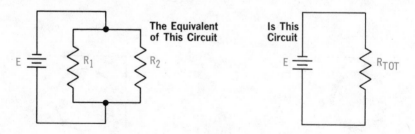

The Equivalent of This Circuit

Is This Circuit

. . . Or the total current can be found by calculating the total resistance of the circuit, and then finding the total current as you would for series circuits

calculating total resistance

In a series circuit, the total resistance is simply the sum of all the individual resistances. The more resistances there are, the greater is the total resistance. It is obvious, therefore, that the total resistance is larger than any of the individual resistances.

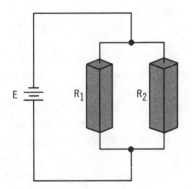

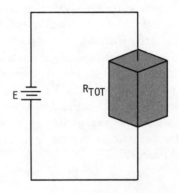

How is the total value of parallel resistances smaller than the individual resistances? This will be clear if you recall that the resistance of a material can be lowered by increasing its cross-sectional area

And this is effectively what is done when resistances are connected in parallel. Each resistance path allows more total current to flow

In a parallel circuit, the relationship between the total resistance and the individual resistances is completely different, and, as a matter of fact, almost opposite. For parallel circuits, the total resistance is *not* the sum of the individual resistances; the *more* resistances there are, the *lower* is the total resistance; and the *total* resistance is *smaller* than any of the individual resistances. The reason for this is that each new branch resistor draws more current from the source to increase the total current. And an increase in total current can only occur from a decrease in total resistance. You can probably see from this that calculating the total resistance for a parallel circuit is quite different than for a series circuit.

There are various ways to find the total resistance of parallel resistances. The best method to use in any particular case usually depends on how many individual resistances there are, and whether or not their values are equal.

equal resistances in parallel

The simplest arrangement of parallel resistances is when two are in parallel, and their values are *equal*. If you think of the two resistances as two equal-sized pieces of the same material, you know that the two together have twice the cross-sectional area of either one. They could, therefore, be replaced in the circuit with one piece of the material having the same length but double the cross-sectional area.

If you glance again at page 2-13, you will remember that if you double the cross-sectional area of a piece of material and do not change its length, you cut its resistance in half. Thus, the total resistance of *two* equal resistances in parallel is *one-half* the value of either resistance. In the same way, the total resistance of *three* equal resistances in parallel is *one-third* the value of any one of the resistances. We can therefore state a general rule for equal resistances in parallel: *The total resistance of any number of parallel resistances all having the same value is equal to the value of one resistance divided by the number of resistances.*

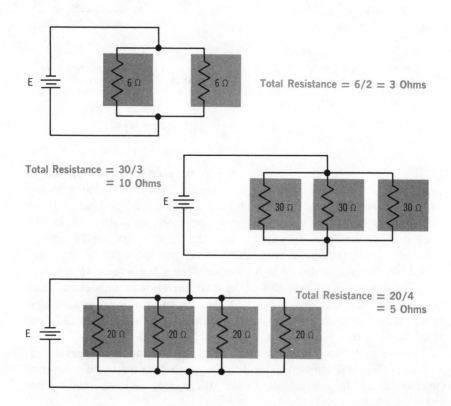

equal resistances in parallel (cont.)

Usually, if the parallel resistances are not equal, this equal-resistance rule cannot be used. However, if the different resistance values are *multiples* of one another, a variation of the equal-resistance rule can be used because any one resistance can be considered as two or more other resistors in parallel. For example, a 4-ohm resistor can be thought of as two 8-ohm resistors in parallel, or three 12-ohm resistors in parallel, or ten 40-ohm resistors in parallel. Therefore, when the parallel resistors have values that are multiples of each other, the lower valued resistors can be thought of as combinations to produce the same effective circuit with equal resistances. Then the equal-resistance rule can be used. This method also applies when the resistor values are not multiples of each other, but can be divided by a common multiple. As an example, 10- and 15-ohm resistors can both be changed to 30-ohm equivalents.

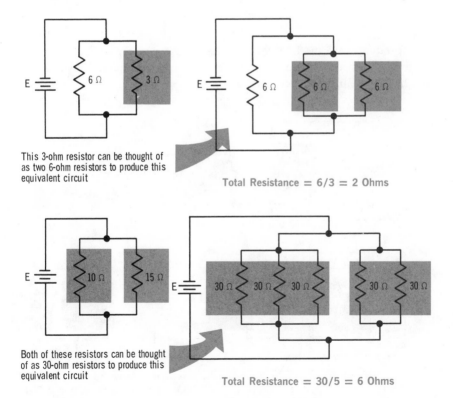

This 3-ohm resistor can be thought of as two 6-ohm resistors to produce this equivalent circuit

Total Resistance = 6/3 = 2 Ohms

Both of these resistors can be thought of as 30-ohm resistors to produce this equivalent circuit

Total Resistance = 30/5 = 6 Ohms

two unequal resistances in parallel

When there are two resistances in parallel, but their values are not the same and cannot be converted to the same multiple, the total resistance cannot be found with the equal-resistance rule. Instead, you must use a method called the *product/sum,* or product over the sum, method. To use this method, you first multiply the values of the two resistances to get their *product.* Then you add the values of the two resistances to get their *sum.* Finally, you divide the product by the sum, and the result is the total resistance. For example, let's use a set of resistor values that were used on page 2-83 to show how this method will produce the same results. If the values of two parallel resistances were 6 ohms and 3 ohms, you first multiply their values ($6 \times 3 = 18$) to get a product of 18. You then add their values ($6 + 3 = 9$) to get a sum of 9. Finally, you divide the product by the sum ($18/9 = 2$) to obtain the total resistance of 2 ohms. As an equation, this method can be expressed as:

$$R_{TOT} = \frac{\text{product}}{\text{sum}} = \frac{R_1 \times R_2}{R_1 + R_2}$$

The product/sum method can also be used for two parallel resistances whose values are the same or have the same multiple. However, this might often cause unnecessary work.

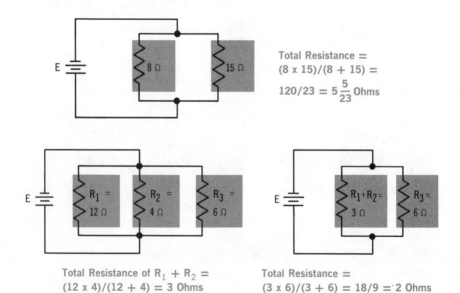

Total Resistance =
(8 x 15)/(8 + 15) =
$120/23 = 5\frac{5}{23}$ Ohms

Total Resistance of $R_1 + R_2$ =
(12 x 4)/(12 + 4) = 3 Ohms

Total Resistance =
(3 x 6)/(3 + 6) = 18/9 = 2 Ohms

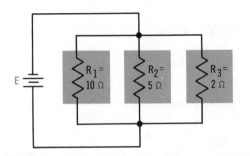

$$R_{TOT} = \cfrac{1}{\cfrac{1}{R_1} + \cfrac{1}{R_2} + \cfrac{1}{R_3}}$$

The reciprocal method can be used to calculate the total resistance of any combination of parallel resistances. In the circuit,

$$R_{TOT} = \cfrac{1}{\cfrac{1}{10} + \cfrac{1}{5} + \cfrac{1}{2}}$$

$$= \cfrac{1}{\cfrac{2}{20} + \cfrac{4}{20} + \cfrac{10}{20}}$$

$$= \cfrac{1}{\cfrac{16}{20}} = \cfrac{20}{16}$$

$$= 1\text{-}1/4 \text{ Ohms}$$

three or more unequal resistances in parallel

The product/sum method you have just learned is a special case of a much more general method for calculating the total resistance of a parallel circuit. The general method can be used for *any number* of resistances, and works whether the resistances are *equal* or *unequal*. The method is actually derived from the equations for Ohm's Law.

This general method is called the *reciprocal method*. The reciprocal of a number is 1 divided by that number. Thus, the reciprocal of 2 is ½, or 0.5; the reciprocal of 4 is ¼, or 0.25; and the reciprocal of 10 is 1/10, or 0.1.

To use the reciprocal method, you first calculate the reciprocal of the value of each resistance. Then add these reciprocals to get the total reciprocal. And finally, find the reciprocal of the total reciprocal. The reciprocal method is described by the equation:

$$R_{TOT} = \cfrac{1}{\cfrac{1}{R_1} + \cfrac{1}{R_2} + \cfrac{1}{R_3} + \ldots + \text{etc.}}$$

calculating total resistance from current and voltage

The various methods you know to calculate the effective resistance of parallel circuits involve only resistance values. However, you can also use Ohm's Law to find the effective resistance of parallel circuits. If the total circuit current and the applied voltage are known, the total resistance can easily be found using Ohm's Law in the form:

$$R_{TOT} = E/I_{TOT}$$

where E is the source voltage and I_{TOT} is the total current.

If only the source voltage is known, the total current can be found by first calculating each of the branch currents, and then adding them. Ohm's Law is then used to find the total resistance.

When the source voltage and total circuit current are both *not* known, Ohm's Law still can be used to find the effective resistance of a parallel circuit. Assume any source voltage you want, and calculate hypothetical branch currents with this voltage. By adding these branch currents, you find the total hypothetical current. Using the assumed voltage and the total current that you found, use the equation $R_{TOT} = E/I_{ASSUMED}$ to find the total resistance.

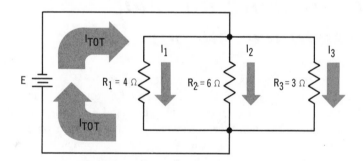

For the circuit, assume a source voltage of 12 volts. The voltage across each branch is, therefore, also assumed to be 12 volts. Therefore:

$$I_{TOT} = E/R_1 + E/R_2 + E/R_3 = 3 + 2 + 4 = 9 \text{ amperes}$$

Then the total resistance can be found by:

$$R_{TOT} = E_{ASSUMED}/I_{TOT} = 12 \text{ volts}/9 \text{ amperes} = 1.33 \text{ ohms}$$

You should obtain the same result by using the reciprocal method.

summary

☐ A parallel circuit is one in which there are one or more points where the current divides and follows different paths. ☐ Parallel loads are connected in such a way that the total current is divided between the loads. ☐ The entire source voltage is dropped across each parallel load. ☐ The separate paths in a parallel circuit are called branches. ☐ The current flowing in each branch is called the branch current. ☐ The total current of a parallel circuit is equal to the sum of the branch currents. ☐ Adding parallel loads to a circuit decreases the total circuit resistance, and thus increases total current.

☐ The total resistance of any number of parallel resistances all having the same value is equal to the value of one resistance divided by the number of resistances. ☐ The total resistance of two unequal resistances in parallel can be found by the product/sum method: $R_{TOT} = (R_1 \times R_2)/(R_1 + R_2)$.

☐ The reciprocal method for finding total resistance can be used for any number of resistances, equal or unequal:

$$R_{TOT} = \frac{1}{1/R_1 + 1/R_2 + 1/R_3 + \ldots +}$$

☐ When the source voltage and total current are not known, the total resistance can be found by assuming a convenient source voltage, solving for the total assumed current, and then solving for the total resistance by $R_{TOT} = E_{ASSUMED} \div I_{TOT}$.

review questions

1. What is a parallel circuit?
2. Find R_{TOT} for six 10-ohm resistors in parallel.
3. Find R_{TOT} for six 10-ohm resistors in series.
4. Doubling the source voltage of a parallel circuit has what effect on the branch currents? On the total circuit resistance?
5. What is the *reciprocal method* equation for finding R_{TOT}?
6. What is the *product/sum* equation for total resistance?
7. What equation should be used for finding the equivalent resistance of two equal resistors in parallel? Of two unequal resistors?
8. When can the reciprocal method be used for finding R_{TOT}?
9. A four-branch parallel circuit has branch currents of 2, 3, 5, and 10 amperes. What is the total current?
10. If the resistance of the 2-ampere branch of Question 9 is 10 ohms, what are the values of the other resistances, and the total resistance?

parallel power sources

Remember that a parallel circuit is a circuit that has more than one path for current flow. So far, the only types of parallel circuits you have studied are those having parallel loads. In these circuits, the current leaving the power source is the total circuit current, and this current is divided between the branches of the circuit.

Another type of parallel circuit is one that has parallel power sources. In this type circuit, each power source supplies *part* of the current that flows through the load. Thus, the current through the load is the *total circuit current,* and the current through each power source is *branch current.*

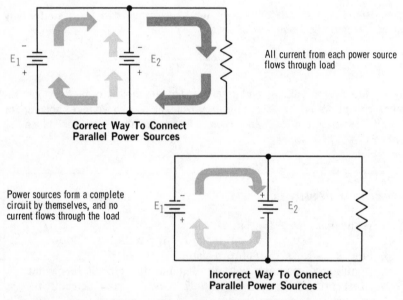

**Correct Way To Connect
Parallel Power Sources**

All current from each power source flows through load

Power sources form a complete circuit by themselves, and no current flows through the load

**Incorrect Way To Connect
Parallel Power Sources**

Each power source connected in parallel supplies part of the total circuit current. This current may be supplied to only one load, or it may be divided among parallel loads

Circuits that have parallel power sources can also have parallel loads. In these circuits, the total current is the *sum* of the currents flowing in the branches.

Only power sources with the *same output voltage* should ever be connected in parallel. If sources with different voltages were put in parallel, some current would flow from the source with the higher voltage into the one with the lower voltage. This would be *wasted* current, since it would not flow through the load.

effect of parallel power sources on current

When two power sources with the same output voltage are connected in parallel, the voltage output of the parallel combination is the *same* as the *individual* sources. There is no increase or decrease in voltage. Therefore, the current through the loads, whether there are one or more of them, is the same as if only one power source was used. Why bother putting power sources in parallel? One answer is that very often power sources cannot supply the *total* current needed by a circuit. But by putting sources in parallel, each only has to supply a *portion* of the circuit current. For example, consider a load of 2 ohms that needs a current of 10 amperes to operate properly. From Ohm's Law, the voltage required:

$$E = IR = 10 \text{ amperes} \times 5 \text{ ohms} = 50 \text{ volts}$$

But suppose that the only 50-volt power sources available can supply 8 amperes at most. In this case, two 50-volt power sources would have to be connected in parallel.

Batteries are often connected in parallel to *extend* their *life*. When current is drawn from a battery, the battery discharges. And the more current that is drawn, the quicker the battery discharges. When batteries are put in parallel, each supplies only a portion of the circuit current. They, therefore, discharge more slowly, and last longer. Power sources and their current limitations are covered in Volume 6.

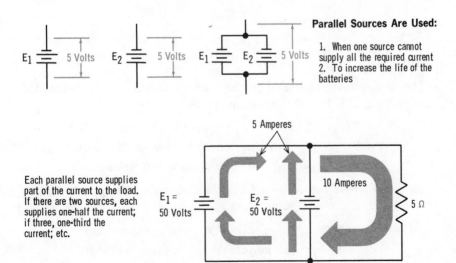

Parallel Sources Are Used:

1. When one source cannot supply all the required current
2. To increase the life of the batteries

Each parallel source supplies part of the current to the load. If there are two sources, each supplies one-half the current; if three, one-third the current; etc.

power consumption

You have learned that in a series circuit, the total power consumed is equal to the sum of the power used by each individual load. You also know that the total power can be calculated directly, using the values of the total circuit current, total circuit resistance, and source voltage, if any two of these values are known. These *same* relationships are true for power in a parallel circuit. It can be calculated *directly* from total current, total resistance, and source voltage; or it can be found *indirectly* by taking the sum of the individual power consumptions of the loads.

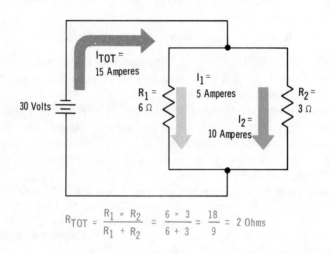

$$R_{TOT} = \frac{R_1 \times R_2}{R_1 + R_2} = \frac{6 \times 3}{6 + 3} = \frac{18}{9} = 2 \text{ Ohms}$$

The total power consumed in the parallel circuit can be calculated directly from either of the three equations:

$$P = EI_{TOT} = 30 \text{ volts} \times 15 \text{ amperes} = 450 \text{ watts}$$
$$P = I^2_{TOT} R_{TOT} = (15 \text{ amperes})^2 \times 2 \text{ ohms} = 450 \text{ watts}$$
$$P = E^2/R_{TOT} = (30 \text{ volts})^2/2 \text{ ohms} = 450 \text{ watts}$$

The total power can also be found by calculating the power used in each branch, and then adding them:

$$P_{R1} = I_1^2 R_1 = (5 \text{ amperes})^2 \times 6 \text{ ohms} = 150 \text{ watts}$$
$$P_{R2} = I_2^2 R_2 = (10 \text{ amperes})^2 \times 3 \text{ ohms} = 300 \text{ watts}$$
$$P_{TOT} = P_{R1} + P_{R2} = 150 + 300 = 450 \text{ watts}$$

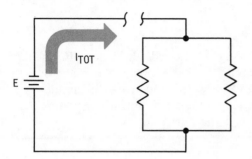

Breaking a parallel circuit where the total current flows opens the entire circuit, stopping all current flow. With no current, there is no power consumed in the circuit and no voltage drops across the loads

Breaking a parallel circuit where branch current flows opens only that branch. Total current and current in the other branches still flow. The values of the total currents, however, will change. The reason is that when a branch is opened, its resistance is no longer part of the circuit, and the total resistance is increased

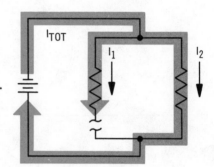

open circuits

If a series circuit is broken at any point, no current flows. The reason for this is that there is only one path for current flow in a series circuit, and that path must be complete or the circuit is open. A parallel circuit, however, has more than one current path. Thus, even if one of the paths is opened, current will still flow in the circuit as long as one or more of the other paths provide a *complete* circuit from the negative terminal of the power source to the positive terminal. This does not mean, however, that you cannot stop current flow in a parallel circuit by opening it at one point. What it does mean is that the behavior of a parallel circuit that is open at some point depends on where the opening, or break, is.

If the circuit is opened at a point through which *total* circuit current flows, the entire circuit is open, and all current flow stops. If, however, it is opened at a point where only a *branch* current flows, then only that branch is open, and current continues to flow in the rest of the circuit. You can see from this that for a fuse to do its job in a parallel circuit, it must be connected at a place where total circuit current flows, or else each branch must have a fuse.

short circuits

When a parallel circuit becomes shorted, the same effects happen that occur when a series circuit is shorted. You remember that these effects include a sudden and very large increase in circuit current, heating of the connecting wires and possible burning of the insulation, and the possible burning out of the power source.

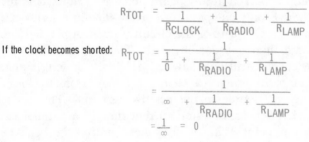

If any load in a parallel circuit becomes shorted, the resistance of the circuit drops to practically zero. The reason is that each load is connected across the power source terminals. This can be shown by the equation:

$$R_{TOT} = \cfrac{1}{\cfrac{1}{R_{CLOCK}} + \cfrac{1}{R_{RADIO}} + \cfrac{1}{R_{LAMP}}}$$

If the clock becomes shorted:
$$R_{TOT} = \cfrac{1}{\cfrac{1}{0} + \cfrac{1}{R_{RADIO}} + \cfrac{1}{R_{LAMP}}}$$

$$= \cfrac{1}{\infty + \cfrac{1}{R_{RADIO}} + \cfrac{1}{R_{LAMP}}}$$

$$= \frac{1}{\infty} = 0$$

Parallel circuits are usually more likely to develop damaging short circuits than are series circuits. The reason for this is that each parallel load is connected *directly* between the power source terminals. And so if any one of the loads becomes shorted, it drops the resistance between the power source terminals to practically zero. But if a series load becomes shorted, the resistance of the other loads in series with it keep the circuit resistance from dropping to zero.

solved problems

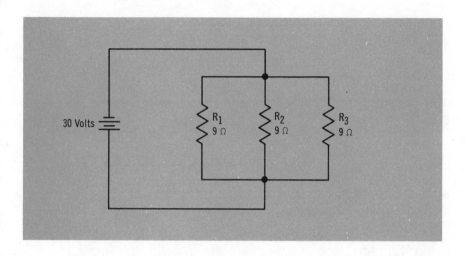

Problem 9.   What is the total current in this circuit?

One way to find the total current would be to calculate the three branch currents and then add their values. However, since the three resistances are equal, it is easier in this case to find the total resistance and then calculate the total current from that. The total resistance of equal parallel resistances is given by:

$$R_{TOT} = \frac{\text{value of one resistance}}{\text{number of resistances}} = \frac{9}{3} = 3 \text{ ohms}$$

The total current can be found using Ohm's Law:

$$I_{TOT} = E/R_{TOT} = 30 \text{ volts}/3 \text{ ohms} = 10 \text{ amperes}$$

Problem 10.   What is the current through R_2?

R_2 is one of the branches in the circuit since it provides a separate path for current. And you know that the current in a branch depends on the resistance of the branch and the voltage across it. So, using Ohm's Law:

$$I_2 = E/R_2 = 30 \text{ volts}/9 \text{ ohms} = 3.33 \text{ amperes}$$

This problem could also have been solved by just looking at the circuit. Since the three branch resistances are equal the total current must divide equally between them. That means one-third of the total current flows through each branch. With the total current being 10 amperes, the branch current through R_2 is one-third of 10, or 3.33 amperes.

solved problems (cont.)

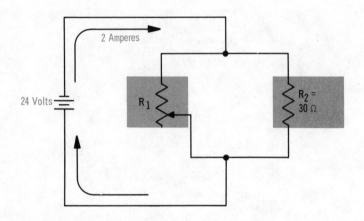

Problem 11. What is the total resistance in the circuit?

Since both branch resistances are not known, you cannot use the reciprocal method or the product/sum method to find the total resistance. But the applied voltage and the total current are known, and so you can use Ohm's Law to calculate the total resistance:

$$R_{TOT} = E/I_{TOT} = 24 \text{ volts}/2 \text{ amperes} = 12 \text{ ohms}$$

Problem 12. If R_1 was set to have a value of 60 ohms, what would the total current be then?

The total resistance of the parallel combination has to be found before the total current can be calculated. There are two unequal resistances, so the product/sum method can be used:

$$R_{TOT} = (R_1 \times R_2)/(R_1 + R_2) = (60 \times 30)/(60 + 30)$$
$$= 1800/90 = 20 \text{ ohms}$$

With the total resistance and the applied voltage known, the total current can then be calculated by Ohm's Law:

$$I_{TOT} = E/R_{TOT} = 24 \text{ volts}/20 \text{ ohms} = 1.2 \text{ amperes}$$

Problem 13. How much of this total current of 1.2 amperes flows through R_2?

The voltage across R_2 is 24 volts and its resistance is 30 ohms. The current through R_2 is, therefore, found by the equation:

$$I = E/R_2 = 24 \text{ volts}/30 \text{ ohms} = 0.8 \text{ ampere}$$

solved problems (cont.)

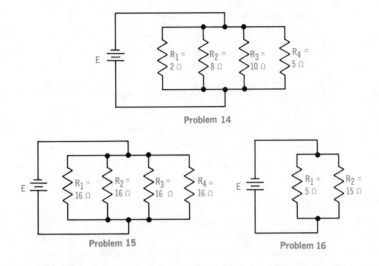

Problem 14

Problem 15

Problem 16

Problem 14. **What is the total resistance of the circuit?**
There are more than two resistances in this parallel combination, so the reciprocal method is used:

$$R_{TOT} = \cfrac{1}{\cfrac{1}{R_1} + \cfrac{1}{R_2} + \cfrac{1}{R_3} + \cfrac{1}{R_4}} = \cfrac{1}{\cfrac{1}{2} + \cfrac{1}{8} + \cfrac{1}{10} + \cfrac{1}{5}}$$

$$= \cfrac{1}{\cfrac{20}{40} + \cfrac{5}{40} + \cfrac{4}{40} + \cfrac{8}{40}} = \cfrac{1}{\cfrac{37}{40}} = 40/37 = 1\text{-}3/37 \text{ ohms}$$

Problem 15. **What is the total resistance of the circuit?**
The reciprocal method could be used for this circuit, but since all the resistances have the same value, it is much easier to use the equal resistance method. Thus,

$$R_{TOT} = \frac{\text{value of one resistance}}{\text{number of resistances}} = \frac{16}{4} = 4 \text{ ohms}$$

Problem 16. **What is the total resistance of the circuit?**
The product/sum method can be used here since there are two resistances with different values:

$$R_{TOT} = (R_1 \times R_2)/(R_1 + R_2) = (5 \times 15)/(5 + 15) = 3.75 \text{ ohms}$$

This problem can also be solved by the equal resistance rule if R_1 is converted to three 15-ohm resistors. Then, there would be a total of four 15-ohm resistors in the circuit, so 15/4 equals 3.75 ohms.

solved problems (cont.)

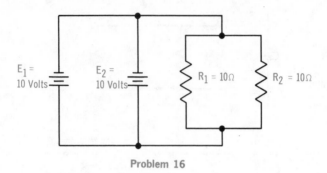

Problem 16

Problem 17. **What is the total current in this circuit?**

By now, you should be able to just look at these two parallel resistances and mentally calculate their total value. They are equal, and so their total value is one-half the value of either, or 5 ohms.

Once you know the total resistance, all you need is the source voltage and you can calculate the total current from Ohm's Law. You remember that if equal power sources are connected in parallel, their combined output voltage is the same as their individual voltages. Thus, the source voltage for this circuit is 10 volts. The total circuit current is, then:

$$I_{TOT} = E/R_{TOT} = 10 \text{ volts}/5 \text{ ohms} = 2 \text{ amperes}$$

And since parallel power sources each supply only a part of the total circuit current, each battery in this circuit is supplying 1 ampere.

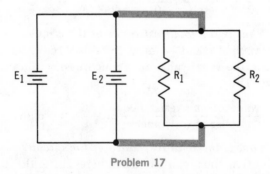

Problem 17

Problem 18. **Where would a fuse have to be connected in the above circuit so that it could open the entire circuit if it blew?**

The fuse would have to be connected at some point where total circuit current flowed. The path of the total current is shown by the colored lines in the circuit.

comparison of series and parallel circuits

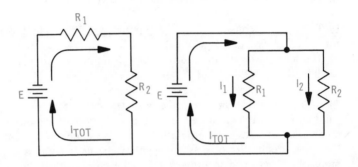

	Series Circuit	Parallel Circuit
Current	There is only one path for current to flow.	There is more than one path for current to flow.
	The current at every point in the circuit is the same.	The total current is equal to the sum of the branch currents.
Voltage	The sum of the voltage drops across the individual loads is equal to the source voltage.	The voltage across each branch is the same as the source voltage.
Resistance	The total resistance is equal to the sum of the individual resistances.	The total resistance is equal to the reciprocal of the sum of the reciprocals of the individual resistances.
Power	The total power consumed is equal to the sum of the power consumptions of the individual loads.	

summary

☐ When power sources are connected in parallel, each source supplies part of the circuit current. ☐ Only power sources with the same output voltage should ever be connected in parallel. ☐ When two power sources with the same output voltage are connected in parallel, the voltage output of the parallel combination is the same as the individual sources.

☐ Total power consumption in a parallel circuit is found the same way as for a series circuit: $P_{TOT} = EI_{TOT}$, $P_{TOT} = I_{TOT}^2/R_{TOT}$, and $P_{TOT} = E^2/R_{TOT}$. ☐ The total power is equal to the sum of the powers consumed in the branch circuits: $P_{TOT} = P_1 + P_2 + P_3 + \ldots + $ etc.

☐ In a parallel circuit, even if one of the current paths is opened, current will still flow in the circuit as long as one or more of the other paths provides a complete circuit. ☐ If a parallel circuit is opened at a point through which the total circuit current flows, the entire circuit is open, and all current flow stops. ☐ For a fuse to protect a parallel circuit, it should be placed in the circuit where total current flows, or else each branch must have a fuse.

---

review questions

1. Why shouldn't power sources having different output voltages be connected in parallel?
2. If three 10-volt batteries are connected in parallel, what is their total output voltage?

For Questions 3 to 8, consider a circuit consisting of three parallel resistors, whose values are 20, 50, and 80 ohms, connected across a 150-volt battery.

3. What is the total resistance of the circuit?
4. What is the total current and the branch currents?
5. What is the total power supplied by the battery, and what is the power consumed by each resistor?
6. Where should a fuse be placed to protect the circuit?
7. Would a 2-ampere fuse protect the battery against a short?
8. If the 20-ohm resistor became open-circuited, what would be the value of the total circuit current?
9. A circuit consists of ten equal-value resistors in parallel. If the total power dissipated by the circuit is 40 watts, and if the circuit is powered by a 10-volt battery, what is the value of each resistor?
10. For the circuit of Question 9, if the battery output falls to 5 volts, what is the total power dissipated by the circuit?

series-parallel circuits

From the material given so far, it should be easy for you to recognize both series and parallel circuits. But there is another type of circuit that has branches, like parallel circuits, and series loads or components,

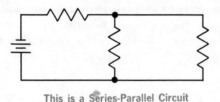

This is a Series-Parallel Circuit

Series-parallel circuits are a combination of both series circuits and parallel circuits. They can be fairly simple and have only a few components, but they can also have many components and be quite complicated

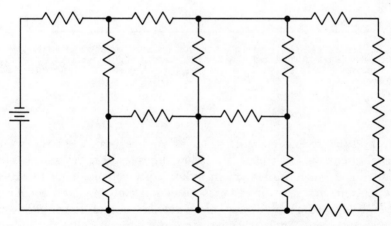

And This is a Series-Parallel Circuit

like series circuits. This is called a *series-parallel* circuit since it is a combination of the others. The methods you will use to analyze series-parallel circuits are mostly combinations of those you already know for series circuits and parallel circuits.

analyzing series-parallel circuits

In any d-c circuit, there are certain basic factors in which you will be interested. From what you have learned about series and parallel circuits, you know that these factors are (1) the total current from the power source and the current in each part of the circuit, (2) the source voltage and the voltage drops across each part of the circuit, and (3) the total resistance and the resistance of each part of the circuit. Once you know these circuit factors, you can easily calculate others, such as total power or the power consumed in some part of the circuit.

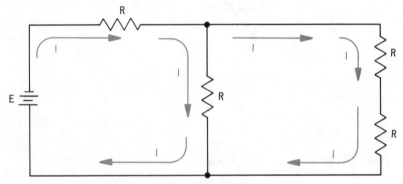

Every branch and every load in a series-parallel circuit has current, voltage and resistance. The circuit has also total current, total voltage, and total resistance. To find any of these, you have to use the rules for series circuits, as well as those for parallel circuits

To find the various currents, voltages, and resistances in series and parallel circuits is fairly easy. You know the rules of series and parallel circuits; and when working with either type, you use only the rules that apply to that type. In a series-parallel circuit, on the other hand, some parts of the circuit are *series-connected* and some parts are *parallel-connected*. Thus, in some parts of a series-parallel circuit, you have to use the rules for series circuits, and in other parts the rules for parallel circuits apply.

You can see then, that before you can analyze or solve a problem involving a series-parallel circuit you have to be able to *recognize* which parts of the circuit are series-connected and which parts are parallel-connected. Sometimes this is obvious, if the circuit is simple. Many times, however, you will have to *redraw* the circuit, putting it into a form that is easier for you to recognize.

redrawing series-parallel circuits

In a series circuit, the current is the same at all points. In a parallel circuit, there are one or more points where the current divides and flows in separate branches. And in a series-parallel circuit, there are both separate branches and series loads. You can see, then, that the easiest way to find out whether a circuit is a series, parallel, or series-parallel circuit is to start at the negative terminal of the power source and *trace* the path of current through the circuit and back to the positive terminal of the power source. If the current does not divide anywhere, it is a series circuit. If the current does divide into separate branches, but there are no series loads, it is a parallel circuit. And if the current divides into separate branches and there are also series loads, it is a series-parallel circuit. When tracing the circuit in this way, remember that there are *two types* of series loads. One type consists of two or more resistances in one branch of the circuit. The other type is any resistance through which the total circuit current flows. You can see these two types in the illustration.

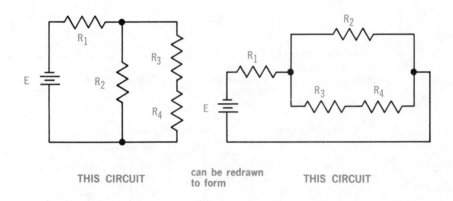

THIS CIRCUIT   can be redrawn to form   THIS CIRCUIT

The analysis of a series-parallel circuit can be often simplified if the circuit is redrawn to that the branches and series loads can be recognized quickly

Often after determining that a circuit is series-parallel, you will find it very helpful to redraw the circuit so that the branches and the series loads are more easily recognized. This will be especially helpful when you have to find the total resistance of the circuit. Examples of how circuits can be redrawn are shown on this and the next page.

redrawing
series-parallel circuits (cont.)

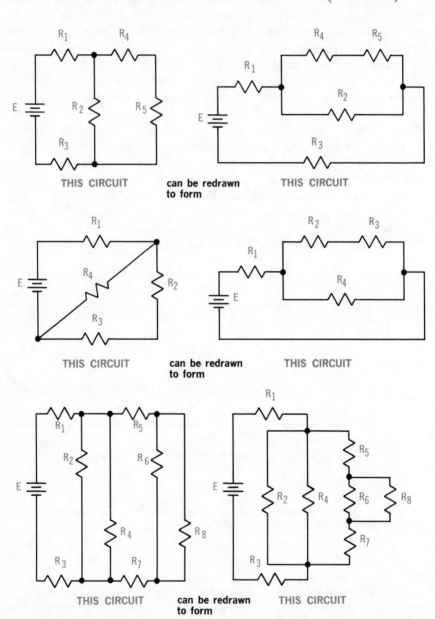

THIS CIRCUIT   can be redrawn to form   THIS CIRCUIT

THIS CIRCUIT   can be redrawn to form   THIS CIRCUIT

THIS CIRCUIT   can be redrawn to form   THIS CIRCUIT

reducing series-parallel circuits

Very often, all that is known about a series-parallel circuit is the applied voltage and the values of the individual resistances. To find the voltage drop across any of the loads or the current in any of the branches, you usually have to know the total circuit current. But to find the total current you have to first know the total resistance of the circuit. To find the total resistance, you reduce the circuit to its *simplest*

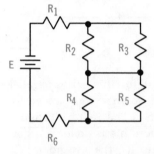

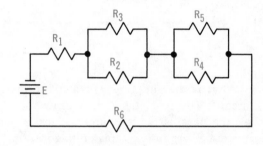

1. If necessary, redraw the circuit so that all parallel combinations of resistances and series resistances are easily recognized

2. For each parallel combination of resistances, calculate its effective resistance

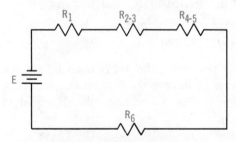

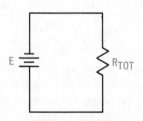

3. Replace each of the parallel combinations with one resistance, whose value is equal to the effective resistance of that combination. This gives you a circuit with all series loads

4. Find the total resistance of this circuit by adding the resistances of all the series loads

form, which is usually *one resistance* that forms a series circuit with the *voltage source*. This simple series circuit has the equivalent resistance of the series-parallel circuit it was derived from, and so also has the same total current. There are four basic steps in reducing a series-parallel circuit.

reducing
series-parallel circuits (cont.)

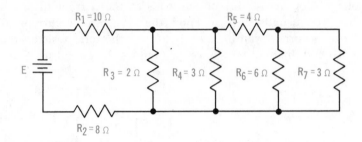

What is the simplest equivalent series circuit for this series-parallel circuit? The first step is to redraw the circuit so that you can easily see the parallel portions and the series portions. (The following figure references are to diagrams on page 2-105.) When this is done, you can see that there are three main branches: one is R_3, the second is R_4, and the third is R_5 and the parallel combination of R_6 and R_7. You have to reduce the three branches so they have one resistance each. (See Fig. A.)

Start by finding the effective resistance of R_6 and R_7, and putting this in the circuit in place of R_6 and R_7 (See Fig. B):

$$R_{6\text{-}7} = (R_6 \times R_7)/(R_6 + R_7) = (6 \times 3)/(6 + 3) = 18/9 = 2 \text{ ohms}$$

You now have two series resistances in the third branch. So you reduce this branch to one resistance by adding the two series resistances and replacing them with a single resistance whose value is equal to their sum. (See Fig. C.)

The three branches now contain one resistance each. These three parallel resistances can be replaced with one resistance, whose value can be found by using the reciprocal method. (See Fig. D.)

$$R_{3\text{-}4\text{-}5\text{-}6\text{-}7} = \cfrac{1}{\cfrac{1}{R_3} + \cfrac{1}{R_4} + \cfrac{1}{R_{5\text{-}6\text{-}7}}} = \cfrac{1}{\cfrac{1}{2} + \cfrac{1}{3} + \cfrac{1}{6}}$$

$$= \cfrac{1}{\cfrac{3}{6} + \cfrac{2}{6} + \cfrac{1}{6}} = \cfrac{1}{\cfrac{6}{6}} = 1 \text{ ohm}$$

The circuit has now been reduced to a series circuit having three series resistances. It can be reduced further to only one resistance by adding the values of the three series resistances. (See Fig. E.)

reducing
series-parallel circuits (cont.)

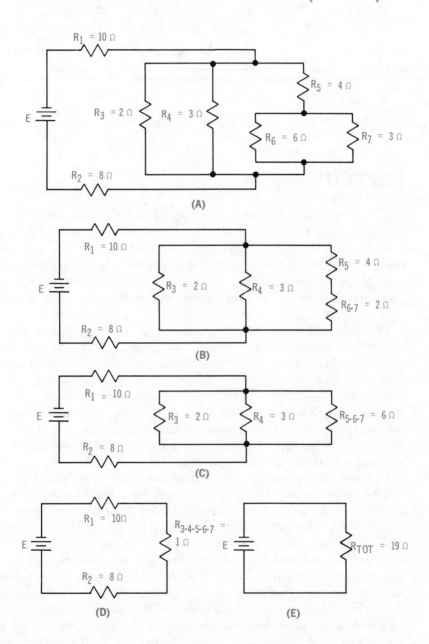

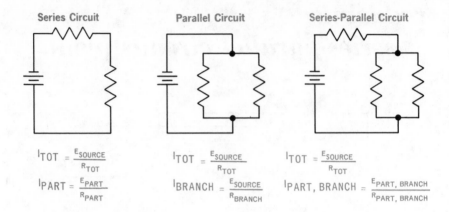

Series Circuit

$$I_{TOT} = \frac{E_{SOURCE}}{R_{TOT}}$$

$$I_{PART} = \frac{E_{PART}}{R_{PART}}$$

Parallel Circuit

$$I_{TOT} = \frac{E_{SOURCE}}{R_{TOT}}$$

$$I_{BRANCH} = \frac{E_{SOURCE}}{R_{BRANCH}}$$

Series-Parallel Circuit

$$I_{TOT} = \frac{E_{SOURCE}}{R_{TOT}}$$

$$I_{PART, BRANCH} = \frac{E_{PART, BRANCH}}{R_{PART, BRANCH}}$$

current

In any d-c circuit, the total current is equal to the power source voltage divided by the total resistance. For series circuits, this is the only current. Therefore, if you know the total current, you also *know* the current through every part of the circuit. In parallel circuits, the current divides and follows more than one path. Therefore, if you just know the total circuit current, you *do not know* the current in every part of the circuit.

Branch currents are usually calculated by applying Ohm's Law to the voltage across the branch and the branch resistance. In series-parallel circuits, the current also divides, following more than one path. So, like parallel circuits, the branch currents have to be found using Ohm's Law. There is an important difference, however. In both cases you use Ohm's Law in the form $I = E/R$. But for parallel circuits, the voltage across every branch in the circuit is the same, and is equal to the source voltage. Therefore, if the source voltage and branch resistance are known, all branch currents can be found.

In series-parallel circuits, the voltage across every branch is usually *not* the same. So, very often, the voltage must be calculated before the branch current can be found. You may think from this that if you calculated all of the circuit voltages first, you could then use them to find all the currents in the circuit. This is *not* the case, though. The approach you will usually use is to first find the total current in the circuit. Then, using the current, you will calculate the voltage across some part or branch of the circuit. With this voltage you will calculate the current through that part or branch. You will then use the current you find to calculate the voltage across some other part or branch. Using this method, you will eventually find all of the currents and voltages in the circuit.

voltage

In a series circuit, the sum of the voltage drops around the circuit equals the applied voltage. And in a parallel circuit, the voltage across each branch is the same as the applied voltage. No such simple relationship exists for a series-parallel circuit between the applied voltage and the voltages throughout the circuit. However, you know that the voltage dropped across any resistance or group of resistances is equal to the current through the resistance times the value of the resistance. This relationship is true in any d-c circuit, whether series, parallel, or series-parallel. And generally, this is the method used to find voltages in a series-parallel circuit.

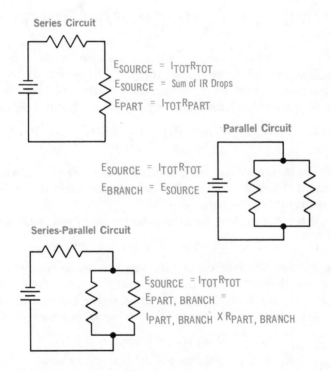

Series Circuit

$$E_{SOURCE} = I_{TOT}R_{TOT}$$
$$E_{SOURCE} = \text{Sum of IR Drops}$$
$$E_{PART} = I_{TOT}R_{PART}$$

Parallel Circuit

$$E_{SOURCE} = I_{TOT}R_{TOT}$$
$$E_{BRANCH} = E_{SOURCE}$$

Series-Parallel Circuit

$$E_{SOURCE} = I_{TOT}R_{TOT}$$
$$E_{PART, BRANCH} =$$
$$I_{PART, BRANCH} \times R_{PART, BRANCH}$$

Remember that you usually cannot calculate all of the currents or all of the voltages in a series-parallel circuit by using only the total current and applied voltage. You have to work around the circuit load by load and branch by branch, finding the current through and voltage across each load or branch before moving on to the next. Of course, as you acquire more experience and practice you will develop shortcuts that will enable you to eliminate some of the work.

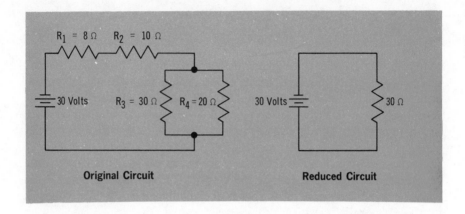

Original Circuit   Reduced Circuit

calculating voltage and current

In the series-parallel circuit, calculate the current through the resistances and voltage drops across them. First reduce the circuit to its simplest form. Start by reducing the parallel combination of R_3 and R_4 to one equivalent resistance by the product/sum method:

$$R_{3\text{-}4} = (R_3 \times R_4)/(R_3 + R_4) = (30 \times 20)/(30 + 20)$$
$$= 600/50 = 12 \text{ ohms}$$

Now the original circuit has been reduced to a series circuit with three resistances: 8, 10, and 12 ohms. The completely reduced circuit, then, has one resistance of 30 ohms.

The total current in this circuit can be found by using Ohm's Law:

$$I = E/R = 30 \text{ volts}/30 \text{ ohms} = 1 \text{ ampere}$$

Referring back to the original circuit, you can see that this 1-ampere current flows through resistances R_1 and R_2, and then separates through R_3 and R_4. Since you know the current through R_1 and R_2, calculate the voltage drops across them using Ohm's Law.

$$E_{R1} = IR_1 = 1 \text{ ampere} \times 8 \text{ ohms} = 8 \text{ volts}$$
$$E_{R2} = IR_2 = 1 \text{ ampere} \times 10 \text{ ohms} = 10 \text{ volts}$$

If 8 volts are dropped across R_1 and 10 volts across R_2, 12 volts remain across the combination of R_3 and R_4. The current through each of these can now be found.

$$I_{R3} = E/R_3 = 12 \text{ volts}/30 \text{ ohms} = 0.4 \text{ ampere}$$
$$I_{R4} = E/R_4 = 12 \text{ volts}/20 \text{ ohms} = 0.6 \text{ ampere}$$

You know that both branch currents must add up to the total circuit current of 1 ampere, so you can check your results by adding them.

solved problems

Problem 19.   *Redraw the circuit so that the series and parallel portions are shown more clearly.*

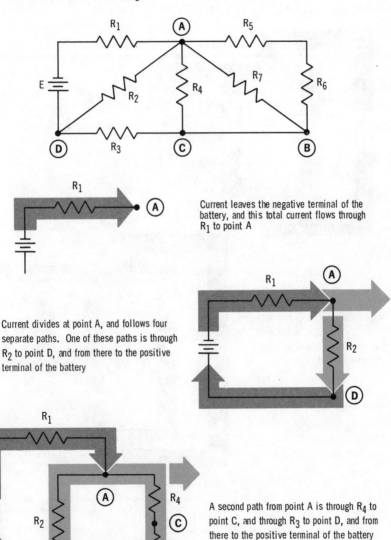

Current leaves the negative terminal of the battery, and this total current flows through R_1 to point A

Current divides at point A, and follows four separate paths. One of these paths is through R_2 to point D, and from there to the positive terminal of the battery

A second path from point A is through R_4 to point C, and through R_3 to point D, and from there to the positive terminal of the battery

The circuit is further simplified on page 2-110.

solved problems (cont.)

The circuit from page 2-109 is further simplified:

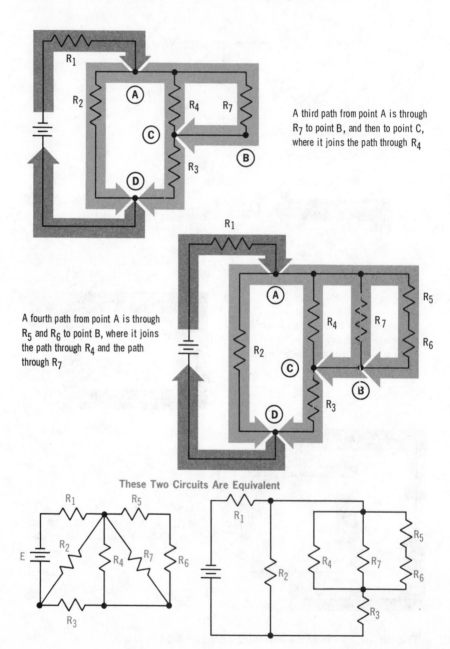

A third path from point A is through R_7 to point B, and then to point C, where it joins the path through R_4

A fourth path from point A is through R_5 and R_6 to point B, where it joins the path through R_4 and the path through R_7

These Two Circuits Are Equivalent

solved problems (cont.)

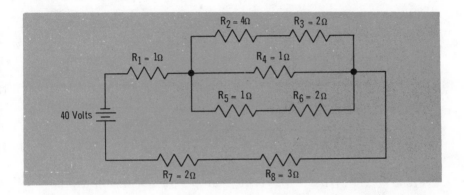

Problem 20. **What is the total current flowing in the circuit?**

To do this, reduce the circuit to only one resistance that is equivalent to the total value of all the series and parallel resistances in the circuit. You can see that there are three branches in the circuit. One of them is through R_2 and R_3, the second is through R_4, and the third is through R_5 and R_6. Each of these branches must first be reduced. R_2 and R_3 are in series; so their total resistance is $4 + 2 = 6$ ohms. R_5 and R_6 are also in series, and their total resistance is, therefore, $1 + 2 = 3$ ohms.

The resistances of the three branches are then 6 ohms, 1 ohm, and 3 ohms, and the effective resistance of the three branches can be found by the reciprocal method for calculating parallel resistances:

$$R_{EFF} = \frac{1}{\dfrac{1}{6} + \dfrac{1}{1} + \dfrac{1}{3}} = \frac{1}{\dfrac{1}{6} + \dfrac{6}{6} + \dfrac{2}{6}}$$

$$= \frac{1}{\dfrac{9}{6}} = 0.667 \text{ ohms}$$

Now replace the parallel resistances with the 0.667-ohm resistance, and the result is a series circuit with four resistances. Since it is a series circuit, the total resistance is 6.667 ohms. This reduces the original circuit to a simple one having 40 volts across 6.667 ohms. The total current, then, is found by Ohm's Law:

$$I = E/R = 40 \text{ volts}/6.667 \text{ ohms} = 6 \text{ amperes}$$

solved problems (cont.)

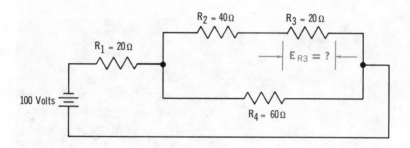

Problem 21. **What is the voltage drop across R₃ in the circuit?**

To find the voltage drop across R_3, you have to know the current through it. The current is not given, and so you must first calculate the total circuit current, and then find how much of the total current flows through R_3.

Start by reducing the circuit to a single equivalent resistance. There are two parallel branches: one is R_2 and R_3, and the other is R_4. You know the total resistance of the R_4 branch, and you find the total of the R_2-R_3 branch by adding the resistances: $40 + 20 = 60$ ohms. Both branches have total resistances of 60 ohms, so the effective resistance of the parallel combination is one-half the value of either branch, or 30 ohms. The resistance of the entire circuit is the 30 ohms plus the 20 ohms of R_1 or 50 ohms. The total circuit current can be found by Ohm's Law:

$$I_{TOT} = E/R_{TOT} = 100 \text{ volts}/50 \text{ ohms} = 2 \text{ amperes}$$

This total current flows through R_1, and drops some voltage across it.

$$E_{R1} = IR_1 = 2 \text{ amperes} \times 20 \text{ ohms} = 40 \text{ volts}$$

This means that the difference between 40 volts and the power source voltage, or 60 volts, must be across the two parallel branches.

With 60 volts across each branch, the current through the branch containing R_2 and R_3 can be found.

$$I = E/R_{2\text{-}3} = 60 \text{ volts}/60 \text{ ohms} = 1 \text{ ampere}$$

This current flows through both R_2 and R_3, since they are in series. You know the current through R_3, and can therefore calculate the voltage drop across it using Ohm's Law:

$$E_{R3} = IR_3 = 1 \text{ ampere} \times 20 \text{ ohms} = 20 \text{ volts}$$

summary

☐ A series-parallel circuit can be identified by tracing the path of current. If the current divides into separate branches, and there are also series loads, the circuit is a series-parallel circuit. Any resistance through which the total current flows is in the series part of the circuit. ☐ Any resistance through which only part of the total current flows is in the parallel part of the circuit. ☐ In a series-parallel circuit, all the laws for series and parallel circuits are obeyed.

☐ Series-parallel circuits are often reduced to their simplest form to solve for electrical quantities. The simplest form is usually one resistance that forms a series circuit with the voltage source. ☐ Four basic steps in reducing series-parallel circuits are: Redraw the circuit, if necessary, so that all parallel combinations of resistances and series resistances are easily recognized. ☐ Calculate the effective resistance of each parallel combination of resistances. ☐ Replace each parallel combination with one resistance equal to the effective resistance of the combination, resulting in a series circuit. ☐ Find the total resistance of the circuit by adding the resistances of all series loads.

☐ The total voltage dropped around a series-parallel circuit equals the source voltage. ☐ The branch currents through a parallel part of a circuit can be found by calculating the voltage drop across the parallel part, and then applying Ohm's Law. This voltage drop is equal to the total current times the series-equivalent resistance of the parallel part. ☐ The power dissipated by each load of a series-parallel circuit is found by the standard power equations. ☐ The total power is equal to the power developed by the equivalent total resistance of the circuit, or to the sum of the powers developed by each load.

review questions

1. Define *series, parallel,* and *series-parallel circuits.*

For Questions 2 to 10, consider the circuit on page 2-104, with all 4-ohm resistors and a 100-volt power source.

2. What is the total circuit resistance?
3. What is the total circuit current?
4. What is the current through R_2?
5. What is the voltage across R_1?
6. What is the voltage across R_3, and across R_7?
7. What is the current through R_4?
8. What is the current through R_6?
9. What is the power consumed by R_5?
10. What is the total power consumed in the circuit?

law of proportionality

You have learned how to apply Ohm's Law in simple and complex circuits to calculate current, voltage, or resistance. For many circuits it might be convenient to reduce resistances and trace current the way the earlier lessons taught. But as you become experienced and adept at the calculations, you will probably notice that the *relative values* of resistances in a circuit will allow some short-cut methods of calculations. The law of proportionality is an example of this.

Take the series circuit on this page. The source voltage is given, and so are both resistor values. Ordinarily, if you want to find the voltage drop across R_1, you would use Ohm's Law to find first the current flow in the circuit, and then again use Ohm's Law to calculate the voltage drop. But the law of proportionality allows you to find the voltage drop across R_1 with only one calculation.

Since the same current is flowing through both resistors, the voltage drop across each is directly proportional to the value of that resistor. If one resistor is twice the value of the other, it will have twice the voltage drop; if it is three times the value, it will have three times the voltage drop; and so on. So a short-cut method is to arrange an equation that will show what percentage the resistance of the resistor in question is of the overall total circuit resistance, and then multiply that percentage by the total voltage to get the voltage drop.

Suppose you wanted to find the voltage drop across R_1. The equation

$$\frac{R_1}{R_1 + R_2}$$

will show the proportion of resistance that R_1 has of the total circuit resistance. If you multiply that by the total voltage, you will find E_1:

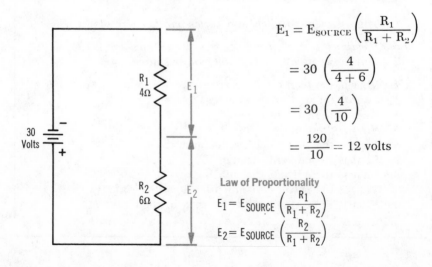

$$E_1 = E_{\text{SOURCE}} \left(\frac{R_1}{R_1 + R_2} \right)$$

$$= 30 \left(\frac{4}{4+6} \right)$$

$$= 30 \left(\frac{4}{10} \right)$$

$$= \frac{120}{10} = 12 \text{ volts}$$

Law of Proportionality

$$E_1 = E_{\text{SOURCE}} \left(\frac{R_1}{R_1 + R_2} \right)$$

$$E_2 = E_{\text{SOURCE}} \left(\frac{R_2}{R_1 + R_2} \right)$$

law of proportionality (cont.)

The voltage across R_2 could be found the same way, with the equation:

$$E_2 = E_{\text{SOURCE}} \left(\frac{R_2}{R_1 + R_2} \right)$$

As a double check on the answer, here is how E_1 would be found with ordinary Ohm's Law:

$$R_{\text{TOT}} = R_1 + R_2 = 4 + 6 = 10 \text{ ohms}$$

$$I = \frac{E_{\text{SOURCE}}}{R_{\text{TOT}}} = \frac{30 \text{ volts}}{10 \text{ ohms}} = 3 \text{ amperes}$$

$$E_1 = I R_1 = 3 \times 4 = 12 \text{ volts}$$

As shown in the drawing on this page, the law of proportionality can also be applied to finding currents in parallel circuits. Since the voltage that will be dropped across two parallel resistors is the same, then the current that flows through any one resistor will be *inversely* proportional to the value of that resistor as compared to the value of the other parallel resistor. This means that if one resistor has twice the value of another, it will have half the current, and so on. Notice that the voltage proportionality in the series circuit was *directly* proportional to the value, whereas here, the current value is *inversely* proportional. Because of this the equation here is slightly different. The value of I_1 depends on the proportionality of R_2 and vice versa. Let's find I_1 in this circuit:

$$I_1 = I_{\text{TOT}} \left(\frac{R_2}{R_1 + R_2} \right)$$

$$= 4 \left(\frac{6}{4 + 6} \right)$$

$$= 4 \left(\frac{6}{10} \right)$$

$$= \frac{24}{10} = 2.4 \text{ amperes}$$

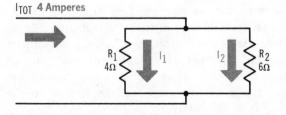

I_{TOT} 4 Amperes

R_1 4Ω I_1 I_2 R_2 6Ω

Law of Proportionality

$$I_1 = I_{\text{TOT}} \left(\frac{R_2}{R_1 + R_2} \right)$$

$$I_2 = I_{\text{TOT}} \left(\frac{R_1}{R_1 + R_2} \right)$$

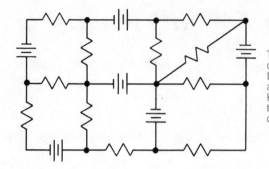

The voltages and currents in this circuit cannot be found by Ohm's Law. There are too many branches and too many power sources. With Kirchhoff's Laws, though, all of the voltages and currents can be calculated

kirchhoff's laws

In all of the circuits examined so far, Ohm's Law has described the relationships between the current, voltage, and resistance. However, all of the circuits covered have been *relatively simple*. There are many circuits that are so complex that they cannot be solved by Ohm's Law. These circuits have *many branches* or *many power sources,* and Ohm's Law would be either impractical or impossible to use on them. Methods are needed, therefore, for solving complex circuits. Any methods used, though, must not violate Ohm's Law, since Ohm's Law is the very basis of d-c circuit theory.

Methods for solving complex circuits have been developed, and are based on the experiments of a German physicist, Gustav Kirchhoff. About 1857, Kirchhoff developed two conclusions as a result of his experiments. These conclusions, known as *Kirchhoff's Laws,* can be stated as follows:

Law No. 1: The sum of the voltage drops around any closed loop is equal to the sum of the emf's in that loop.

Law No. 2: The current arriving at any junction point in a circuit is equal to the current leaving that point.

In using Kirchhoff's Laws, it makes no difference which is called the first law and which is called the second. Since, in use, law No. 1 above is usually applied first, we will call it Kirchhoff's first law. Law No. 2 will, therefore, be his second law.

These two laws may seem obvious to you based on what you already know about circuit theory. In spite of their apparent simplicity, though, they are powerful tools when it comes to solving difficult and complex circuits. Although the laws themselves are simple, the mathematics for applying them becomes more difficult as the circuits become more complex. Because of this, the discussion here will be limited to the use of the laws for solving only the relatively minor complex circuits.

kirchhoff's voltage law

Kirchhoff's first law is also known as his *voltage law*. You will often see it written in many different forms, but no matter what form it is in, it is expressing the same fact. It gives the relationship between the *voltage drops* around any closed loop in a circuit and the voltage sources in that loop. The *totals* of these two quantities are *always equal*. This can be given in equation form as: $\Sigma\,E_{\text{SOURCE}} = \Sigma\,IR$, where the symbol Σ, which is the Greek letter *sigma*, means "the sum of."

These Circuits ARE Loops

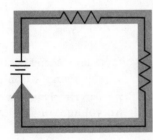 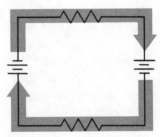

Kirchhoff's Voltage Law can only be applied to closed loops. A closed loop must satisfy two conditions:

1. **It must have one or more voltage sources**
2. **It must have a complete path for current flow from any point, around the loop, and back to that point**

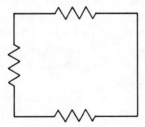 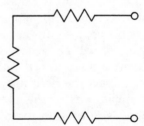

These Circuits ARE NOT Loops

You will recall that in a simple series circuit, the sum of the voltage drops is equal to the applied voltage. This is actually Kirchhoff's voltage law applied to the simplest possible case, that is, where there is only one loop and one voltage source in that loop.

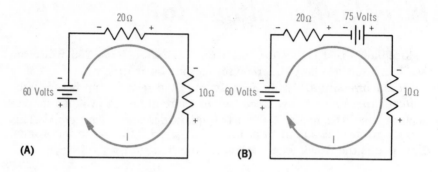

applying kirchhoff's voltage law

For a simple series circuit, Kirchhoff's voltage law corresponds to Ohm's Law. To find the current in circuit (A) by using Kirchhoff's voltage law, simply use the equation: $\Sigma E_{\text{SOURCE}} = \Sigma IR$. There is only one voltage source, or emf, in the loop, and two voltage, or IR, drops. So the equation becomes:

$$60 = 20I + 10I$$
$$60 = 30I$$
$$I = 60/30 = 2 \text{ amperes}$$

In the above problem, the direction of the current flow was known before the problem was solved. When there is more than one voltage source, the direction of current might not be known. In this case, you assume a direction at the beginning of the problem. All the sources that would aid the current in this assumed direction are then positive, and all that would oppose current in this direction are negative. The answer to the problem will be positive if you assumed the correct direction of current flow, and negative if you assumed the wrong direction. In either case, you will get the right magnitude of current.

For example, what is the current in circuit (B)? If you assume that the current is flowing in the direction shown, the equation for Kirchhoff's voltage law is

$$\Sigma E_{\text{SOURCE}} = \Sigma IR$$
$$60 - 75 = 20I + 10I$$
$$-15 = 30I$$
$$I = -15/30 = -0.5 \text{ ampere}$$

The result is negative, and so the current is actually 0.5 ampere in the direction opposite to what we assumed.

kirchhoff's current law

Kirchhoff's second law is called his *current law*. Like the voltage law, it too is often stated in different ways. No matter how it is stated, though, its meaning does not change. And the law is: at any junction point in a circuit, the current *arriving* is *equal* to the current leaving. This should be obvious to you from what you learned in Volume 1. Current cannot *collect* or *build up* at a point. For every electron that arrives at a point, one must leave. If this was not so, potential would build up, and current would eventually stop when the potential equaled that of the power source. Thus, if 10 amperes of current arrives at a junction that has two paths leading away from it, the 10 amperes will divide among the two paths, but the total 10 amperes must leave the junction. You are already familiar with the most obvious application of Kirchhoff's second law from parallel circuits. That is, that the sum of the branch currents is equal to the total current entering the branches as well as to the total current leaving the branches.

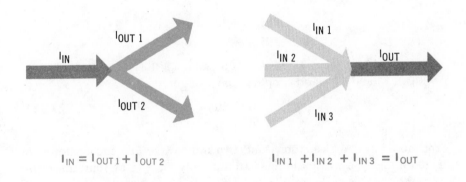

$$I_{IN} = I_{OUT\,1} + I_{OUT\,2} \qquad\qquad I_{IN\,1} + I_{IN\,2} + I_{IN\,3} = I_{OUT}$$

Kirchhoff's current law states that current cannot collect at a point. The current that leaves a point must be equal to the current that enters a point. Thus, if you assign a positive polarity to current entering a point and a negative polarity to the current leaving a point, the algebraic sum of the currents at any point is zero:

$$\Sigma I_{IN} - \Sigma I_{OUT} = 0$$

$$\text{or} \quad \Sigma I_{IN} = \Sigma I_{OUT}$$

Normally, Kirchhoff's current law is not used by itself, but together with the voltage law in solving a problem. This is shown on the following pages.

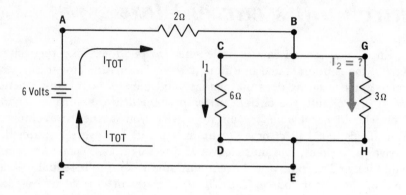

applying kirchhoff's laws

Find the current through the 3-ohm resistance in the circuit using Kirchhoff's Laws. There are two loops in this circuit: one is ABCDEFA, and the other is ABGHEFA. First apply Kirchhoff's voltage law to both loops:

$$2I_{TOT} + 6I_1 = 6 \qquad (1)$$

$$2I_{TOT} + 3I_2 = 6 \qquad (2)$$

Now, since $I_{TOT} = I_1 + I_2$, if you substitute $(I_1 + I_2)$ in place of I_{TOT} in Eqs. (1) and (2), and simplify, you will get:

$$8I_1 + 2I_2 = 6 \qquad (3)$$

$$2I_1 + 5I_2 = 6 \qquad (4)$$

You now have two equations and two unknowns, and must eliminate I_1 to find I_2. One way to do this is to multiply Eq. (4) by four, and subtract Eq. (3) from the result:

$$8I_1 + 20I_2 = 24$$
$$- (8I_1 + 2I_2 = 6)$$
$$\overline{18I_2 = 18}$$

You now have an equation with only I_2, the current you are looking for. So the current I_2 through the 3-ohm resistor is

$$18I_2 = 18$$

$$I_2 = 18/18 = 1 \text{ ampere}$$

You could have solved this problem by just using Ohm's Law, but it was solved by Kirchhoff's Laws to show you the techniques used in solving complex circuits when Ohm's Law cannot be used.

the principle of superposition

When there is more than one power source in a circuit or loop, the current is affected by *each* of the sources. You have examined two ways of finding the current in such cases. One is to determine the combined voltage of the sources, and then use Ohm's Law to find the current. The other way was to use Kirchhoff's voltage law around the loop. A third method that you can use is based on the fact that the current at any point is the *sum* of the currents caused by *each* of the power sources. Therefore, if you can calculate the current that would exist if there was only one power source, and do this for each individual source, the sum of these currents will be the total current with all the sources acting in the circuit. This is called the principle of *superposition*. There are four steps involved in applying the principle:

1. Replace all power sources with a short circuit, except one, and assume a direction of current flow.
2. Calculate the current you want with the one source in the circuit.
3. Do this for each power source in the circuit.
4. Add the individual currents you found. Currents in the direction of the assumed current flow are positive. Those in the opposite direction are negative. If the total current turns out negative, the assumed direction was wrong.

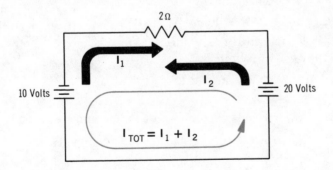

For the circuit shown, we assume a direction for the total current. Thus,

$$I_1 = E_1/R = -10 \text{ volts}/2 \text{ ohms} = -5 \text{ amperes}$$

$$I_2 = E_2/R = 20 \text{ volts}/2 \text{ ohms} = 10 \text{ amperes}$$

$$I_{TOT} = I_1 + I_2 = -5 + 10 = 5 \text{ amperes}$$

an example of superposition

Problem: Find the currents through each resistor, and calculate their voltage drops.

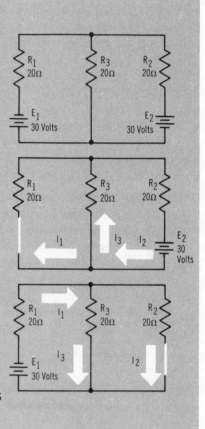

$$R_{TOT} = R_2 + \frac{R_1 R_3}{R_1 + R_3}$$

$$= 20 + \frac{20 \times 20}{20 + 20} = 20 + \frac{400}{40}$$

$$= 20 + 10 = 30 \text{ ohms}$$

$$I_2 = E_2/R_{TOT} = 30/30 = 1 \text{ ampere}$$

Since $R_1 = R_3$, the current $I_1 = 0.5$ ampere and $I_3 = 0.5$ ampere.

$$R_{TOT} = R_1 + \frac{R_2 R_3}{R_2 + R_3}$$

$$= 20 + \frac{20 \times 20}{20 + 20} = 20 + \frac{400}{40}$$

$$= 20 + 10 = 30 \text{ ohms}$$

$$I_1 = E_1/R_{TOT} = 30/30 = 1 \text{ ampere}$$

Since $R_3 = R_2$, the current $I_3 = -0.5$ ampere and $I_2 = 0.5$ ampere.

Answers: $I_{1TOT} = I_1 + I_1 = (0.5) + (1.0) = 1.5$ amperes

$I_{2TOT} = I_2 + I_2 = (1.0) + (0.5) = 1.5$ amperes

$I_{3TOT} = I_3 + I_3 = (0.5) + (-0.5) = \text{zero current}$

Since the currents I_1 and I_2 are in the same direction in both calculations, they are both positive currents. I_3, however, is in opposite directions in both calculations; so one of them is negative.

$$E_1 = I_1 R_1 = 1.5 \times 20 = 30 \text{ volts}$$

$$E_2 = I_2 R_2 = 1.5 \times 20 = 30 \text{ volts}$$

$$E_3 = I_3 R_3 = (0) \times 20 = \text{zero volts}$$

thevenin's theorem

Earlier in the book you learned how to reduce a complex circuit to its simplest form, so that you could make the basic Ohm's Law calculations that could be applied to the actual overall circuit. Often, in many circuits, the circuit data is needed only for a load resistor at the output, and it would be convenient if all of the calculations did not have to be made. Also, if the load resistor value still were not decided, it would be impossible to make any calculations because each different value of load resistor would change the currents and voltage drops throughout the circuit.

A late 19th century scientist, Leon Thevenin, pondered this problem and developed a theory that any complex circuit with a two terminal output could be reduced to a simple *equivalent circuit* across whose output any load resistor would function in the same way as the original circuit. *Thevenin's theorem* states that the equivalent circuit need contain only an equivalent voltage source, $E_{THEVENIN}$, and an equivalent series resistance, $R_{THEVENIN}$.

As a proof that this is possible, take the circuit shown on this page. We want to find the current in R_L. First disconnect R_L. The rest of the circuit is to be thevenized. If you connect a voltmeter across the output terminals, you will measure Thevenin's equivalent voltage, E_{THEV}. You can also calculate it using Ohm's Law or the law of proportionality. You will find that $E_{THEV} = 4.5$ volts. Next connect an ammeter to read Thevenin's current. Since the ammeter will short out R_2, only R_1 will be left in the circuit to give $I_{THEV} = E/R = 6/2 = 3$ amperes.

Now to find Thevenin's equivalent resistance, divide E_{THEV} by I_{THEV}, and you will get 1.5 ohms, R_{THEV}. So Thevenin's equivalent circuit will be as shown, with E_{THEV} in series with R_{THEV} and R_L. Simple Ohm's Law can now be used to find the current through R_L, no matter what value you decide to use. This same problem was solved with Kirchhoff's Laws a few pages earlier.

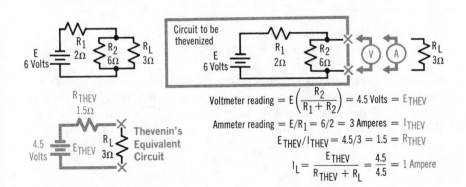

Voltmeter reading $= E\left(\dfrac{R_2}{R_1 + R_2}\right) = 4.5$ Volts $= E_{THEV}$

Ammeter reading $= E/R_1 = 6/2 = 3$ Amperes $= I_{THEV}$

$E_{THEV}/I_{THEV} = 4.5/3 = 1.5 = R_{THEV}$

$I_L = \dfrac{E_{THEV}}{R_{THEV} + R_L} = \dfrac{4.5}{4.5} = 1$ Ampere

thevenizing

Actually, the discussion on the previous page only showed the proof that Thevenin's theorem works. To actually thevenize a circuit, you would not use a voltmeter or ammeter, nor would you make *all* calculations that were made. Thevenin's theorem requires that you:

1. First determine E_{THEV} across the output terminals with R_L disconnected.

2. *Look back into* the output terminals to see what equivalent resistance R_{THEV} exists with the source voltage *shorted out.*

E_{THEV} is found just the way it was on the previous page, either by Ohm's Law or by the law of proportionality.

To find R_{THEV}, when the source voltage is shorted out, resistors R_1 and R_2 become connected in parallel when viewed from the output terminals.

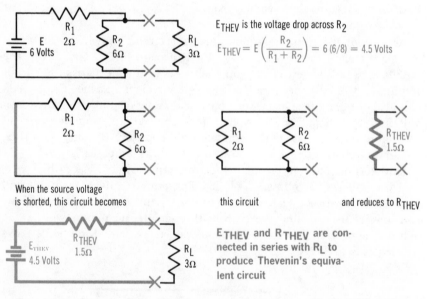

E_{THEV} is the voltage drop across R_2

$$E_{THEV} = E\left(\frac{R_2}{R_1 + R_2}\right) = 6\,(6/8) = 4.5\text{ Volts}$$

When the source voltage is shorted, this circuit becomes

this circuit

and reduces to R_{THEV}

E_{THEV} and R_{THEV} are connected in series with R_L to produce Thevenin's equivalent circuit

So, using the parallel resistance equation, you will find that R_{THEV} = 1.5 ohms, and Thevenin's equivalent circuit is the same as previously shown.

To show you the real value of Thevenin's equivalent circuit, you can now test a variety of different R_L values, and determine the current, voltage, and power with simple Ohm's Law in each case. If you used the original circuit to do this, complex computations would have to be repeated for each value of R_L tested.

a sample problem

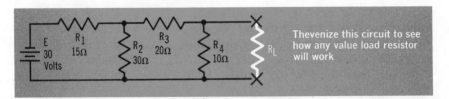

Thevenize this circuit to see how any value load resistor will work

First find the voltage drop across R_4, which will be E_{THEV}

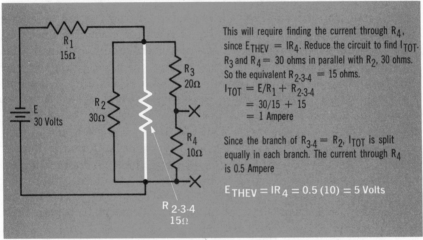

This will require finding the current through R_4, since $E_{THEV} = IR_4$. Reduce the circuit to find I_{TOT}. R_3 and $R_4 = 30$ ohms in parallel with R_2, 30 ohms. So the equivalent $R_{2\text{-}3\text{-}4} = 15$ ohms.

$$I_{TOT} = E/R_1 + R_{2\text{-}3\text{-}4}$$
$$= 30/15 + 15$$
$$= 1 \text{ Ampere}$$

Since the branch of $R_{3\text{-}4} = R_2$, I_{TOT} is split equally in each branch. The current through R_4 is 0.5 Ampere

$$E_{THEV} = IR_4 = 0.5\,(10) = 5 \text{ Volts}$$

Now short out the source voltage and find Thevenin's equivalent resistance, R_{THEV}, looking back from the output terminals

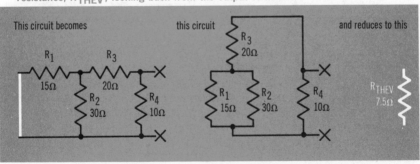

This circuit becomes   this circuit   and reduces to this

Draw Thevenin's equivalent circuit

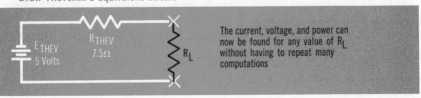

The current, voltage, and power can now be found for any value of R_L without having to repeat many computations

norton's theorem

Norton's theorem is another method of reducing a circuit to a simple equivalent. It is similar in concept to Thevenin's theorem, but differs in that it uses an equivalent constant current source instead of an equivalent voltage source. It also uses an equivalent resistance, but whereas Thevenin's resistance was put in series with the load, Norton's resistance is placed in parallel with the load.

To show how Norton's theorem accomplishes the same as Thevenin's, let's use the same basic circuit.

First find Norton's constant current source, I_N, which will be shown as an arrow in a circle showing the current direction. Disconnect R_L from the output terminals, and connect a short-circuit wire across the terminals. Calculate the current flowing into the wire. Since the wire shorts out resistor R_2, then resistor R_1, which is 2 ohms, is the only resistance across the 6-volt source. So Norton's current, I_N, is 3 amperes.

Next, find Norton's equivalent resistance, R_N. This is found exactly the same way as Thevenin's resistance, by shorting out the voltage source and reducing the circuit looking back into the output terminals.

Once I_N and R_N are found, the equivalent circuit is drawn with the constant current source feeding R_N in *parallel* with the loads. Notice that the current flow in R_L and the drop across R_L are the same in both Thevenin's and Norton's equivalent circuits; Ohm's Law shows them to be 1 ampere and 3 volts in both circuits.

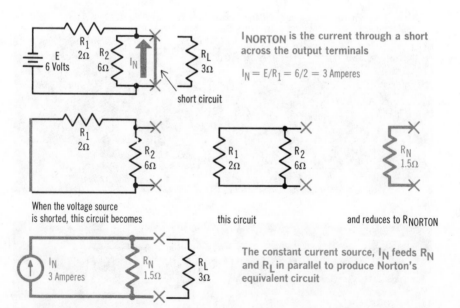

I_{NORTON} is the current through a short across the output terminals

$I_N = E/R_1 = 6/2 = 3$ Amperes

short circuit

When the voltage source is shorted, this circuit becomes

this circuit

and reduces to R_{NORTON}

The constant current source, I_N feeds R_N and R_L in parallel to produce Norton's equivalent circuit

internal resistance
of power sources

You will remember from page 2-19 that every power source has some *internal resistance* that opposes current flow. Normally this resistance is very small and has little effect on circuit operation. For this reason, the internal resistance of the power source has been neglected in all circuits throughout this volume. When the internal resistance of a source must be considered in a circuit, it is usually represented as a resistance in series with the source. And for the most part, this is the effect it has on a circuit: that of an *additional resistance* in the circuit in series with the power source.

Every resistance in a circuit has voltage dropped across it when current flows. Since the internal resistance of the source is inside the source, its voltage drop is also *internal*. This internal drop *subtracts* from the source output voltage. And since the amount of any voltage drop follows the equation $E = IR$, the higher the internal resistance of a source or the more current it conducts, the greater will be its internal voltage drop, and the lower will be its output voltage.

For some power sources, the current output is limited by the internal voltage drop. If you try to draw more than a certain amount of current from these sources, the increased current causes an increase in the internal voltage drop, which lowers the output voltage and, therefore, decreases the current.

When power sources are connected in parallel, their internal resistances are also in parallel. So their effective internal resistance is lower than that of any of the individual sources.

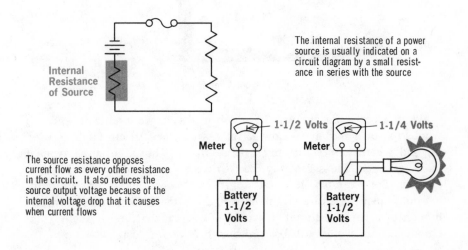

Internal
Resistance
of Source

The internal resistance of a power source is usually indicated on a circuit diagram by a small resistance in series with the source

The source resistance opposes current flow as every other resistance in the circuit. It also reduces the source output voltage because of the internal voltage drop that it causes when current flows

1-1/2 Volts   1-1/4 Volts

Meter   Meter

Battery
1-1/2
Volts

Battery
1-1/2
Volts

d-c circuit failures

For a circuit to operate correctly, *every part* has to do its job. The *power source* has to supply the required voltage, the *connecting wires* have to provide low-resistance connections between the circuit parts without overheating or short-circuiting, and the *loads* have to do their job without drawing too much current from the power source. Auxiliary devices such as switches also have to operate properly. You can see, then, that if any part in a circuit goes bad, or *fails*, the entire circuit fails, since it cannot operate in the way it was designed.

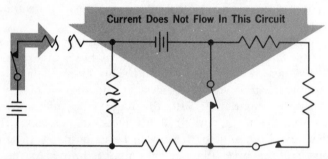

When an electrical part opens, this does not necessarily mean that the circuit is open. Likewise, when a part becomes shorted, it does not always short the entire circuit

The exact effect that a faulty part has on a circuit depends on what the part is and how it fails, as well as the type of circuit and the location of the part in the circuit

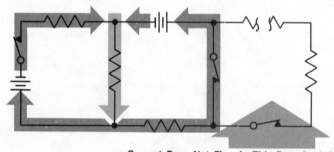

Current Does Not Flow In This Branch of the Circuit

When a part fails, the way in which it affects the circuit depends on what type of a part it is, how it fails, what type of circuit it is used in, and where it is connected in the circuit. Thus, the same type of part, failing in the same way, can cause different effects in different circuits. For example, if a resistor became open in a series circuit, it would open the entire circuit. But if the same resistor opened in one branch of a parallel circuit, it would only open that one branch. Total circuit current would still flow, although it would decrease.

power source failures

A common failure in d-c circuits is failure of the power source. Sources can fail by losing their output entirely, or by having their output voltage decrease to a value less than what is considered normal.

In the field of electricity, the two most common d-c power sources are the d-c generator and the battery. D-c generators can fail because of some mechanical defect within the generator, or because of electrical troubles, such as open or shorted wires.

The most common type of battery failure is a decrease in output voltage caused by the discharging of the battery. Any battery gradually discharges as current is drawn from it over a period of time. When it has discharged to the point where its output voltage is less than that needed to operate the circuit, the battery either has to be recharged or replaced. Batteries can also develop internal short circuits, and then must be replaced. However, this does not happen frequently. D-c generators and batteries are described in detail in Volume 6.

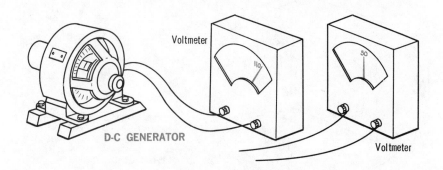

D-c generators and batteries fail whether by losing their outputs completely, or by having their output voltages decrease considerably

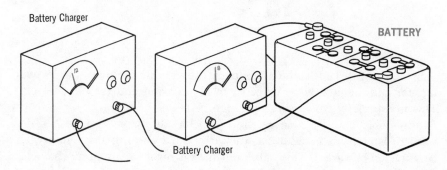

resistor failures

Resistors are one of the most common causes of failure in a circuit. This is not because they are exceptionally delicate or prone to damage, but rather because so many are used. In electric circuits, resistors fail in two main ways: (1) they burn out and *open,* and (2) their resistance value *changes.* Resistors can change in value so that their resistances go up or down. If the value increases greatly, it will act like an open resistor. If the value decreases greatly, it will act like a shorted resistor. These failures are usually the result of the *heat* generated in the resistor by the current. If the heat is not severe, but is generated for a long period of time, the resistor can change in value, while if the heat is very severe, the resistor will more likely burn out in a short time. Often, a resistor failure is caused by some *other failure* in the circuit which results in an increase in circuit current. The increased current then overheats the resistor and causes it to fail.

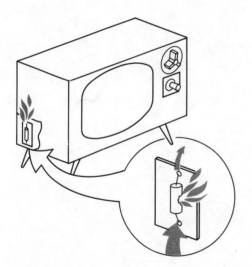

Most resistor failures result from the heat generated in the resistor by the current flow. This heat can cause resistors to change in value, as well as to burn out and become open

Since most resistor failures are caused by too much heat, a defective resistor can frequently be detected by its charred and discolored look. This is especially true for composition resistors.

Adjustable and variable resistors develop the same troubles as do fixed resistors, but in addition, they can fail by losing their ability to be adjusted or varied. This can happen as a result of some mechanical defect, such as a broken sliding contact, or because of an electrical trouble, like a dirty contact.

failure of other loads

Resistors are not the only type of load that can fail in a circuit. Actually, any device used as a load can develop trouble and cause a circuit failure. The loads that use some sort of *heating element* to accomplish their function are a frequent cause of failure. Examples of this type of load are toasters, irons, electric heaters, and lamps. During their normal operation, these loads experience severe heating, which causes the material in the element to *expand*. When the current is turned off, the element cools, and *contracts* to its original size. The amount of this expansion and contraction is not great, but it occurs each time the device is turned on and off. This alternate expansion and contraction *fatigues* the material, and eventually the element *breaks* and opens the circuit.

Loads that use a heating element usually fail because of the alternate heating and cooling of the element when the circuit is turned on and off. The heating and cooling cause expansion and contraction of the element material, and eventually the element breaks

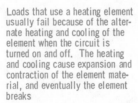

failure of auxiliary devices

The two circuit auxiliary devices you are familiar with are the *switch* and the *fuse*. Both of these can fail and affect circuit operation.

Since the purpose of a switch is to open and close a circuit, any switch must have two positions: one to allow current to flow, and the other to stop current flow. The switch is operated mechanically to one position or the other. If a switch *sticks* in one position or *breaks* so that it stays in one position, its basic switching action is lost. It then keeps the circuit either closed or open at all times. Another type of switch failure occurs when the switch *contacts* become dirty. You remember that a switch should have no effect on a circuit when it is closed. It should, therefore, have almost zero resistance. But if the contacts are dirty, the dirt can act as an *additional resistance* in the circuit and cause a decrease in circuit current.

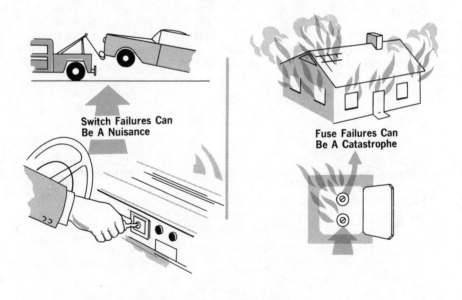

Switch Failures Can
Be A Nuisance

Fuse Failures Can
Be A Catastrophe

Fuses do not fail very often, but if they do, the results can be catastrophic. If, due to imperfections in the fuse material, a fuse should blow at a *lower* current level than it is supposed to, the circuit will be opened *unnecessarily*. This is inconvenient, but not serious. But, if a fuse does not blow when it should, the *high damaging currents* that it is supposed to prevent will flow in the circuit. These currents can burn out the power source or load, and possibly even start a fire.

summary

☐ Proportionality is an easier way of finding series voltage drops and parallel currents. ☐ Kirchhoff's voltage law is: The sum of the voltage drops around any closed loop is equal to the algebraic sum of the emf's in that loop. Mathematically, $\Sigma E_{SOURCE} = \Sigma IR$. ☐ Kirchhoff's current law is: The current arriving at any point in a circuit is equal to the current leaving that point. Mathematically, $\Sigma I_{IN} = \Sigma I_{OUT}$.

☐ In using Kirchhoff's voltage law, assume a direction for the loop current and traverse the loop in the direction of the assumed current, returning to the starting point. ☐ The emf's in the circuit are positive if they tend to aid the current in the assumed direction, and are negative if they tend to oppose the current. ☐ If the value of current found using Kirchhoff's laws is negative, the assumed direction of current flow was wrong.

☐ The principle of superposition is used to solve for current when there is more than one power source. The current through a particular load is found by adding the currents due to each source. The current caused by each source is calculated by replacing the other sources by their internal resistance, or by a short circuit if the internal resistance is negligible. ☐ Thevenin's and Norton's theorems are used to produce simple equivalent circuits. ☐ Power sources fail by a loss or decrease of output voltage. ☐ Resistors can fail by burning out and opening, or by changing value. ☐ Wires can fail by breaking. ☐Switches can fail by locking in one position, or by having dirty contacts. ☐ Fuses can fail by opening prematurely or failing to open.

review questions

For Questions 1 to 5, consider circuit B on page 2-118.

1. Solve for the current using the principle of superposition.
2. Which end of the 20-ohm resistor is negative?
3. What is the voltage drop across each resistor?
4. What is the total power supplied by the two batteries?
5. Answer Questions 1 to 3 with the 75-volt battery terminals reversed.

For Questions 6 to 10, consider the circuit on page 2-121.

6. Solve for the current using Kirchhoff's voltage law.
7. What is the polarity of the voltage drop across the resistor?
8. What is the value of the voltage drop across the resistor?
9. What is the power consumed by the resistor?
10. Answer Questions 6 to 9 with the 20-volt battery terminals reversed.

HAYDEN BOOKS

A Division of Howard W. Sams & Company

Enhance your study of electricity with the Electricity Series, edited by Harry Mileaf. Beginning with the fundamentals of electricity and atomic theory, each of the seven books covers a given area of knowledge taught in incremental steps so that each volume prepares the student for the next one. The series provides complete coverage from general principles to mechanical energy. All titles listed below are the Revised, Second Edition.

ELECTRICITY SERIES

Electricity One	General Principles & Applications	45945	$12.95
Electricity Two	Electric Circuits	45946	$12.95
Electricity Three	Alternating Current	45947	$12.95
Electricity Four	LCR Circuits	45948	$12.95
Electricity Five	Testing Equipment	45949	$12.95
Electricity Six	Power Sources	45950	$12.95
Electricity Seven	Electric Motors	45951	$12.95
All seven individual volumes in a complete set		45944	$69.65
Volumes 1-4 bound into one hardbound book		45919	$32.95
Volumes 1-7 bound into one hardbound book		45952	$42.95

Mr. Mileaf continues his coverage of fundamentals to electronics with the Electronics Series, seven volumes which completely span the scope of the technology. From the concept of the electronic signal to auxiliary circuits and antennas, you'll learn all the necessary theory and practical applications. All titles listed below are the Revised, Second Edition.

ELECTRONICS SERIES

Electronics One	Electronic Signals	45954	$12.95
Electronics Two	Transmitters & Receivers	45955	$12.95
Electronics Three	Tubes & Diodes	45956	$12.95
Electronics Four	Semiconductors	45957	$12.95
Electronics Five	Power Supplies	45958	$12.95
Electronics Six	Oscillators	45959	$12.95
Electronics Seven	Auxiliary Circuits	45960	$12.95
All seven individual volumes in a complete set		45953	$69.65
Volumes 1-7 bound into one hardbound book		45961	$42.95

To order call 800-428-SAMS

index